Digital Signal Processing

Digital Signal Processing

Steve White
DeVry Institute of Technology
Phoenix, AZ

THOMSON LEARNING

Africa • Australia • Canada • Denmark • Japan • Mexico • New Zealand • Philippines
Puerto Rico • Singapore • Spain • United Kingdom • United States

NOTICE TO THE READER

Publisher does not warrant or guarantee any of the products described herein or perform any independent analysis in connection with any of the product information contained herein. Publisher does not assume, and expressly disclaims, any obligation to obtain and include information other than that provided to it by the manufacturer.

The reader is expressly warned to consider and adopt all safety precautions that might be indicated by the activities herein and to avoid all potential hazards. By following the instructions contained herein, the reader willingly assumes all risks in connection with such instructions.

The Publisher makes no representation or warranties of any kind, including but not limited to, the warranties of fitness for particular purpose or merchantability, nor are any such representations implied with respect to the material set forth herein, and the publisher takes no responsibility with respect to such material. The publisher shall not be liable for any special, consequential, or exemplary damages resulting, in whole or part, from the readers' use of, or reliance upon, this material.

Delmar Staff:
Publisher: Alar Elken
Executive Editor: Sandy Clark
Acquisitions Editor: Gregory L. Clayton
Developmental Editor: Michelle Ruelos Cannistraci
Editorial Assistant: Jennifer Thompson
Executive Marketing Manager: Maura Theriault

Channel Manager: Mona Caron
Executive Production Manager: Mary Ellen Black
Production Manager: Larry Main
Senior Project Editor: Christopher Chien
Art/Design Coordinator: David Arsenault
Marketing Coordinator: Paula Collins
Technology Project Manager: Tom Smith

COPYRIGHT © 2000
Delmar, a division of Thomson Learning, Inc. The Thomson Learning™ is a trademark used herein under license.

Printed in the United States of America
2 3 4 5 6 7 8 9 10 XXX 05 04 03 02 01 00

For more information, contact Delmar, 3 Columbia Circle, PO Box 15015, Albany, NY 12212-0515; or find us on the World Wide Web at http://www.delmar.com

International Division List

Asia
Thomson Learning
60 Albert Street, #15-01
Albert Complex
Singapore 189969
Tel: 65 336 6411
Fax: 65 336 7411

Japan:
Thomson Learning
Palaceside Building 5F
1-1-1 Hitotsubashi, Chiyoda-ku
Tokyo 100 0003 Japan
Tel: 813 5218 6544
Fax: 813 5218 6551

Australia/New Zealand:
Nelson/Thomson Learning
102 Dodds Street
South Melbourne, Victoria 3205
Australia
Tel: 61 39 685 4111
Fax: 61 39 685 4199

UK/Europe/Middle East
Thomson Learning
Berkshire House
168-173 High Holborn
London
WC1V 7AA United Kingdom
Tel: 44 171 497 1422
Fax: 44 171 497 1426

Latin America:
Thomson Learning
Seneca, 53
Colonia Polanco
11560 Mexico D.F. Mexico
Tel: 525-281-2906
Fax: 525-281-2656

Canada:
Nelson/Thomson Learning
1120 Birchmount Road
Scarborough, Ontario
Canada M1K 5G4
Tel: 416-752-9100
Fax: 416-752-8102

Spain:
Thomson Learning
Calle Magallanes, 25
28015-MADRID
ESPANA
Tel: 34 91 446 33 50
Fax: 34 91 445 62 18

ALL RIGHTS RESERVED. No part of this work covered by the copyright hereon may be reproduced or used in any form or by any means—graphic, electronic, or mechanical, including photocopying, recording, taping, Web distribution or information storage and retrieval systems—without the written permission of the publisher.

For permission to use material from this text or product contact us by Tel (800) 730-2214; Fax (800) 730-2215; www.thomsonrights.com

Library of Congress Cataloging-in-Publication Data

ISBN 0-7668-1531-5

Contents

Preface ix

Chapter 1. Introduction to Digital Signal Processing and Digital Filtering 1
   1.1 Introduction 1
   1.2 Historical Perspective 1
   1.3 Simple Examples of Digital Signal Processing 2
   1.4 The Common DSP Equation 5
   1.5 What the DSP Equation Shows 6

Chapter 2. Effect of Signal Sampling 9
   2.1 Periodic Sampling of a Cosine Signal 9
   2.2 Periodicity of Any DSP System Frequency Response 11
   2.3 Aliasing and Nyquist Limit 15
   2.4 Anti-aliasing Filters 16
   2.5 The Nyquist Limit and DSP Output Periodicity by Mathematical Means 18

Chapter 3. Digital Filter Specifications 25
   3.1 Introduction to Filter Gain, Loss, dB, and Graphical Filter Specifications 25
   3.2 The Lowpass Digital Filter Specification 28
   3.3 The Highpass Digital Filter Specification 30
   3.4 The Bandpass Digital Filter Specification 31
   3.5 The Bandstop Digital Filter Specification 33
   3.6 Alternate Graphical Specifications 35

| Chapter | 4. | z-Transforms | 39 |

- 4.1 The Need for z-Transforms of the DSP Equation . . 39
- 4.2 The Definition of the z-Transform and Its Use 40
- 4.3 Derivation of Necessary z-Transform Pairs 43
- 4.4 Derivation of the Major z-Transform Property Using Algebra 46

Chapter 5. The z-Transform of the DSP Equation 55

- 5.1 The Transformation of the Modified DSP Equation 55
- 5.2 The Transfer Function of a Digital Filter 59

Chapter 6. Frequency Response of Digital Filters and DSP Systems 67

- 6.1 The Euler Equation from Trigonometry 67
- 6.2 Frequency Scaling 69
- 6.3 Computing the DSP Frequency Response 70

Chapter 7. IIR Filter Design 81

- 7.1 Review of Four Basic Analog Filter Approximations 82
- 7.2 The Impulse Invariant IIR Filter 84
- 7.3 The Step Invariant IIR Filter 91
- 7.4 The Bilinear Transform (BLT) Filter 94

Chapter 8. Digital Filter and DSP Stability 111

- 8.1 Introduction to Stability 111
- 8.2 The Z-plane Unit Circle 113
- 8.3 Other Properties Using the Z-plane 120

Chapter 9. Filter Coefficient Precision 129

- 9.1 Introduction into Computer Numeric Precision . . 129
- 9.2 Development of Equations for Precision Effects . . 131
- 9.3 Computing DSP Coefficient Word Length Effects . 139

Chapter 10. FIR Filter Design 145

- 10.1 Introduction to the FIR Filter 146
- 10.2 The General FIR Coefficient Equation 147
- 10.3 The Basic Solutions of the Coefficient Equation . . 150
- 10.4 Use of the Basic Solutions 154
- 10.5 The Causal (Real Time) and Noncausal Filter Coefficients 156

Chapter 11. Windows for FIR Filters **165**
    11.1 The Gibbs Effect 165
    11.2 Several Windows 167
    11.3 Non-Windowing Approaches 175

Chapter 12. Practical Digital Filter Considerations **187**
    12.1 FIR versus IIR Digital Filters 188
    12.2 Effects of Analog to Digital Converter Number of Bits 191
    12.3 Fixed Point Math versus Floating Point for a DSP Chip 194
    12.4 Realization Forms for Filters 196

Chapter 13. Digital Integration **207**
    13.1 Introduction to Digital Integration 207
    13.2 Digital Integration of Known Signals 209
    13.3 Digital Integration for Differential Equation Solution 214

Appendix A: Laplace Transform Tables **223**
Appendix B: Entering a Mathcad Program **227**

Index **229**

Preface

Purpose and Scope

Most digital signal processing (DSP) texts are written at the postgraduate level, but not for any technical reason. This text is designed for a one-semester course for students at the junior or senior college level. The student should have a good background in algebra and analog signal processing (ASP). A background in first-semester calculus is helpful but not essential to understanding and using most of the material. Using the material in this text, the student will be able to understand the basic concepts and characteristics of DSP systems, read and communicate on DSP subjects, and analyze and design digital filters.

This text emphasizes the major use of DSP, digital filtering. Subjects applicable to all DSP are pointed out, and a chapter on using DSP for signal integration and equation solution is included. For a course in digital filtering, only Chapters 1 through 12 are needed; Chapter 13, on digital integration, shows how easily the preceding material can be applied to other DSP tasks.

The text uses concepts the student has learned in algebra and in ASP to develop the basic characteristics and concepts of digital filters and DSP systems. Chapter 1 introduces all DSP systems as coding of a simple equation called a difference equation on a computer or DSP chip. Many of the concepts that follow are developed mathematically but illustrated graphically to show the student the intuitive basis for the results.

Tables of values, examples, and figures illustrate the current ideas in the text that the student can follow and verify. Applications present simplified uses of digital filtering in the real world. Students will apply concepts learned throughout the text to solve sample problems encountered in industry. Chapter 7 includes an application using IIR filters. Chapter 11 includes an application using FIR filters. At the end of each chapter, a Self-Test includes problems with answers, so that students can apply and check their knowledge. Another set of problems is included without solutions. Most of the problems can be solved using a scientific calculator. To solve for the many coefficients of a filter difference equation or its frequency response, a mathematical package such as Mathcad makes calculation and plotting much easier.

Pedagogical Features

Use of Mathcad

Considerable thought went into choosing between Mathcad and Matlab to implement the analysis and design equations. Both have additional packages for digital filtering that can be used by the students following this text. For students at the level of this text, however, Mathcad has advantages. Since in Mathcad the equations look like the mathematical equations in this or any text, no time is lost in learning to use Mathcad. The major advantages of using Mathcad are listed below:

- The equations in a Mathcad program look like any mathematical equations typed using an equation editor. This format makes it easier to use, understand, and modify the Mathcad program.

- In a Mathcad program all values that were ever computed are always available for output. This does not hold true in Matlab, if the values were intermediate values in a computation.

- To see a Mathcad value after it has been computed, simply type its name followed by an equal sign.

- It is very easy in Mathcad to add text to equations, plots, and data.

Features

Equal emphasis is given to both infinite impulse filters (IIR) and finite impulse filters (FIR), since both are used in industry; certain segments use one or the other almost exclusively. The effects of digital precision are considered in terms of the number of bits used for analog to digital conversion and computer or DSP chip coefficient precision. Chapter 12 covers the practical aspects of implementing the design of FIR or IIR filters.

The equations for designing an FIR filter are derived from the ideal graphical specifications by using first-semester calculus instead of by finding the Fourier series of the periodic frequency response of a digital system. Both methods yield the same result, but the author has seen how much confusion results from having students apply the Fourier series to a frequency spectrum, when they have been taught to use it to get the frequency spectrum from a periodic time signal.

Functional notation for a sampled time function is used here, instead of the sequence notation used in many texts. Students are familiar with functional notation from algebra; it directly relates the sampled signal value to the input and output signals of digital filters to the corresponding time function. As students are familiar with coding on a processor, the added notation of sequences to describe the processor operations on samples is not needed.

The material in this text has been developed from lecture notes used during two semesters of teaching DSP at the DeVry Institute of Technology at the Phoenix campus. The material was taught successfully to both junior and senior level students, whose comments and suggestions were helpful in achieving the final form of the lectures.

Organization

Chapter 1 introduces and defines DSP and digital filtering, as well as showing the basic DSP equation and system diagram. Chapter 2 mathematically describes a sampled signal and examines the important characteristics of sampled signals so that quantitative results can be obtained. Chapter 3 then uses some of the characteristics from Chapter 2 to determine the graphical specifications of digital filters in a manner analogous to analog graphical specifications.

In order to quantitatively evaluate the effects of DSP coding, the filter as well as the signal must be mathematically described. This is done in Chapter 4 by defining and using z-transforms, which are just algebraic transformations of input and output signals, as opposed to the calculus transformations of analog signals using the Laplace transformations. Using z-transformed signals leads to the mathematical description of the process working on the signals. In Chapter 5 the z-transform is applied to the DSP difference equation to get the transfer function of the corresponding digital filter, which is the mathematical description of the filtering process. Then in Chapter 6 the transfer function is evaluated at different frequencies to show how to get the frequency response of any DSP difference equation.

Chapter 7 shows how to design IIR digital filters, which approximate analog filters. Several methods are shown and examples given. Each method has its advantages and disadvantages. The step-invariant method is included; even though it is only one of many methods of digital filtering, it is a significant method in the related fields of digital control and simulation. Chapter 8 shows how to use the transfer function of the DSP system equation to determine its stability and other filter characteristics. In Chapter 9, the transfer function of a digital filter is used to show the effects of numeric precision (the number of bits used to code the filter coefficients) on the filter characteristics.

Chapter 10 then introduces FIR filters, which have no equivalent analog filters. The simple calculus equation for the coefficients is solved to yield algebraic equations for each of the four basic ideal filters with no (or linear) phase. Several examples are given for using the algebraic equations to get the FIR coefficients. The direct application of the algebraic coefficient equations leads to FIR filter difference equations that display the Gibbs effect. The solution is to modify the coefficients slightly, a practice called windowing; several methods and examples are given in Chapter 11.

Chapter 12 covers several practical digital filter design considerations, such as choosing an IIR or an FIR filter, and the effects on the system output of the number of ADC and DAC bits used. This is different from the choice of the number of bits used to encode the filter coefficients, as discussed in Chapter 9. A discussion on fixed point versus floating point numbers is included. The various ways to mathematically represent the difference equation of a filter are presented; called realizations, they have different effects on the actual filter characteristics such as memory used, speed, accuracy, and stability.

Chapter 13 was added to present another DSP topic, that of digital integration. This is a much different subject than digital filtering, but it uses the same difference equation, with the coefficients found in another way. The subject of digital integration is much neglected in DSP texts, but the students use it almost daily in running simulation programs and mathematical packages to solve differential equations. This chapter, a brief introduction, illustrates the pitfalls inherent in this topic. While for digital filtering the BLT method is superior to the step-invariant method, in this chapter a simple solution of a differential equation shows that the method analogous to the step-invariant is superior to the method analogous to the BLT method.

Acknowledgments

The author and Delmar Publishers would like to thank the following reviewers: Dion Benes, Richard Henderson, Seyed Muhammed Jalali, and Martha Keul from DeVry Institute of Technology.

chapter 1

Introduction to Digital Signal Processing and Digital Filtering

1.1 Introduction

Digital signal processing (DSP) refers to anything that can be done to a signal using code on a computer or DSP chip. To reduce certain sinusoidal frequency components in a signal in amplitude, **digital filtering** is done. One may want to obtain the integral of a signal. If the signal comes from a tachometer, the integral gives the position. If the signal is noisy, then filtering the signal to reduce the amplitudes of the noise frequencies improves signal quality. For example, noise may occur from wind or rain at an outdoor music presentation. Filtering out sinusoidal components of the signal that occur at frequencies that cannot be produced by the music itself results in recording the music with little wind and rain noise. Sometimes the signal is corrupted not by noise, but by other signal frequencies that are of no present interest. If the signal is an electronic measurement of a brain wave obtained by using probes applied externally to the head, other electronic signals are picked up by the probes, but the physician may be interested only in signals occurring at a particular frequency. By using digital filtering, the signals of interest only can be presented to the physician.

1.2 Historical Perspective

Originally signal processing was done only on **analog or continuous time signals** using **analog signal processing (ASP)**. Until the late 1950s digital computers were not commercially available. When they did become commercially available they were large and expensive, and they were used to simulate the performance of analog signal processing to judge its effectiveness. These simulations, however, led to digital processor code that simulated or performed nearly the same task on samples of the signals that the analog systems did on the signal. After a while it was realized that the simulation coding of the analog system was actually a DSP system that worked on samples of the input and output at discrete time intervals.

But to implement signal processing digitally instead of using analog systems was still out of the question. The first problem was that an analog input signal had to be represented as a sequence of samples of the signal, which were then converted to the computer's numerical representation. The same process would have to be applied in reverse to the output of the digitally processed signal. The second problem was that because the processing was done on very large, slow, and expensive computers, practical real-time processing between samples of the signal was impossible. Finally, as we will see in Chapter 9, even if digital processing could be done quickly enough between input samples in order to adequately represent the input signal, high sample rates require more bits of precision than slower ones.

The development of faster, cheaper, and smaller input signal samplers (ADCs) and output converters from digital data to analog data (DACs) began to make real-time DSP practical. Also, the processors were becoming smaller, faster, and cheaper and used more bits. Real-time replacements for analog systems may be just as small, cheap, and accurate and be able to process at a sample rate adequate for many analog signals.

However, testing and modification of the coding for DSP systems led to DSP systems that have no analog signal processing equivalents, yet sometimes perform the signal processing better than the DSP coding developed to replace analog systems. For digital filtering, these processing methods are discussed in Chapters 10 and 11.

1.3   Simple Examples of Digital Signal Processing

Digital signal processing entails anything that can be done to a signal using coding on a computer or DSP chip. This includes digital filtering of signals as well as digital integration and digital correlation of signals. This text concentrates on constant rate digital filtering, with references to where the material is applicable to DSP in general. At the end of the text the techniques developed for digital filtering will be used for the DSP task of integration to show how the concepts and techniques are not limited to digital filtering.

The concepts are very simple. A signal is sampled in time at a constant rate in order to input its magnitude value at periodic intervals into the computer. The sample value of the analog magnitude is converted into a binary number. The sampling and conversion are done with an **analog to digital converter (ADC)**. Now the computer code can work on the signal. The computer code computes an output value, which is converted to an analog magnitude from a binary number and then held constant until a new output is computed to replace it. This is done by a **digital to analog converter (DAC)**. The basic DSP system described here is shown in Figure 1.1.

To illustrate the concept of DSP and to see where more study and analysis are needed, let's look at a few simple things that can be done to a sampled signal. If a signal is sampled every T seconds by an ADC and in the computer the sample value is just multiplied by a constant and then sent to the DAC, you have a digital amplifier. The gain of the amplifier is equal to the coded value of the constant. The following equation describes this digital amplifier, where x is the current input sample value from the ADC and y is the corresponding computer output to the DAC.

$$y = ax \quad \text{A simple digital amplifier}$$

If the sampled value of the input signal is multiplied by T, you have computed an approximation to the area under the signal between samples, as long as the signal doesn't change too much between samples. If this value is added to the previous input sample multiplied by T, you

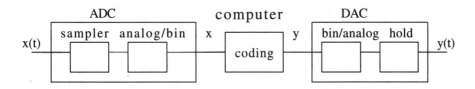

Figure 1.1. Basic DSP system

have approximated the area under the signal over two sample times. This could be repeated endlessly to approximate the area under the signal from when sampling started, as shown in Figure 1.2. The area under a signal or function is its integral. Thus you have performed very simple digital integration using the current sample of the input multiplied by T and then adding the result to the previous output. This process is described by the following equations after two input samples (the -1 subscripts indicate they are previous input or output values).

$$y_{-1} = Tx_{-1} \quad \text{Simple digital integration after one input sample (the previous sample)}$$

$$y = y_{-1} + Tx \quad \text{Simple digital integration after two input samples (the current sample)}$$

By using looping, such as a "While" or "For" loop, the preceding equations could be repeated endlessly by looping about one equation.

If the current input sample value is multiplied by one-half and added to half the previous sample value of the input, a current change in the input signal is reduced, while if the signal is changing slowly the output is very close to the input, since it is just the sum of two half values. Thus the computer is doing a very simple lowpass filtering of the input signal. This simple process is represented by the following equation. The result of using this equation on a string of input samples from the ADC is the input to the DAC shown in Table 1.1. As can be seen, the results, y, are smoothed or lowpass filtered versions of x, the ADC output.

$$y = 0.5x + 0.5x_{-1} \quad \text{Simple digital lowpass filtering}$$

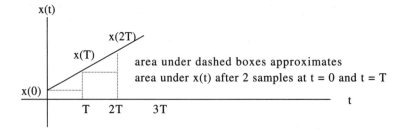

Figure 1.2. Example of digital integration

Introduction to Digital Signal Processing and Digital Filtering

Table 1.1
Example of digital filtering (smoothing)

ADC sample time	0	T	2T	3T	4T
ADC output x	1.2	0.7	1.4	1.1	0.6
DAC input y	xxxxxx	0.95	1.05	1.25	0.85

1.4 The Common DSP Equation

The simple DSP examples just discussed were carried out using some input sample values stored in the computer or received currently from the ADC, multiplying them by appropriate constants, and summing the results. Sometimes the previous output values are multiplied by appropriate constants and also added to the first sum to give a new output, as was done in the digital integration example. Almost all digital signal processing by a computer involves adding the signal input sample just obtained, multiplied by a constant, to the sum of a few previous input samples, each multiplied by their corresponding constants, and sometimes adding all of this to a few previous outputs, each multiplied by their constants, to obtain a new output. This leads to the common equation used for almost all DSP:

$$y = (b_{-1}y_{-1} + \cdots + b_{-m}y_{-m}) + (ax + a_{-1}x_{-1} + \cdots + a_{-n}x_{-n}) \quad \text{(Equation 1.1)}$$

In Equation 1.1, the xs are the sampled input values, the ys are the output samples going to a DAC. The subscripts indicate how many previous sample periods ago are referred to. The as and bs are just constants stored in the computer or DSP chip. A flowchart showing how Equation 1.1 might be implemented by code in the computer shown in Figure 1.1 is given in Figure 1.3.

It may seem strange that almost all DSP tasks are carried out by solving the preceding equation each time a new value of x is input from the ADC, but you must remember that all a computer can do mathematically is add, subtract, multiply, and divide; which is just what this equation requires. If you choose any values for any of the a and b constants and repeat the equation for every new input sample from an ADC, you will be doing DSP! But what DSP have you done and how well? The answers to these questions and more will be given in the rest of this text.

Digital Signal Processing

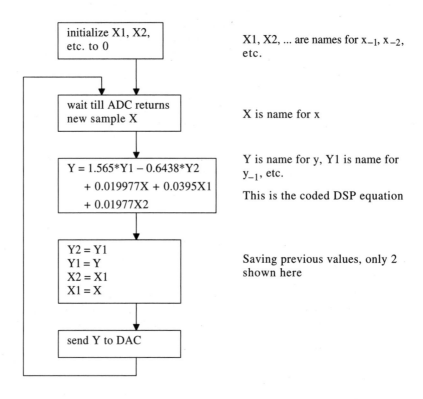

Figure 1.3. Flowchart using the common DSP equation

1.5 What the DSP Equation Shows

The common DSP equation will be used to show that many DSP questions need further study if one is to understand digital signal processing and do analysis or design of a DSP system. These questions include the following:

- How do you choose the *a* and *b* coefficients to perform a specific DSP task, such as doing second-order lowpass Butterworth filtering?

- How many coefficients are needed, and what is the effect of using fewer than required?

- Are the *b* coefficients always needed, and what is the effect if they are not used?

- The *a* and *b* coefficients are represented as binary numbers in the computer; how many bits should be used to meet the filter specifications?

- The *x* values are sample values of the input signal; how often should the signal be sampled?

- What is the effect of different sample rates, and does the filter coding need to be changed if the sample rate changes?

- How many bits should be used in the ADC and DAC to obtain a specific precision?

The answers to these questions and how they are obtained are subjects of the following chapters of this text. In order to fully make use of this text, the student should have a background in college algebra, trigonometry, first-semester calculus, analog filtering, and AC circuits. The only *required* background is in algebra and analog filtering; the others will increase the speed of learning and give a deeper understanding of the subject.

chapter 2

Effect of Signal Sampling

Introduction

In this chapter we examine the effects of sampling on signals and DSP systems. All DSP input signals are sampled, usually at equal intervals of time, in order to input numbers representative of the signal into a computer or DSP chip. We need to determine the effect of this sampling on the signal, as it produces unexpected and critical side effects; these need to be understood before effective filter design can be carried out. In order to simplify the demonstration of the effects of sampling, we will use an analog input signal composed of a single cosine wave. This signal will illustrate and quantitatively show the effects of sampling an analog signal by means of the ADC.

2.1 Periodic Sampling of a Cosine Signal

All signals worked on by DSP systems must be sampled at discrete values of time in order to be input into a DSP chip or computer. This sampling and its conversion to binary values is done by the ADC. This periodic sampling creates very special signal and DSP system characteristics. These characteristics are used in the specification and design of digital filters and DSP systems. In order to see and analyze these characteristics, we will look at the effects of periodic sampling on a cosine signal at different frequencies. We will often refer to signals at, above, or below a certain frequency, rather than to sinusoids with frequencies at, above, or below those of sinusoids at a certain frequency. This shorthand English is used almost universally in industry and the literature. As an example, "reducing the frequencies above 100 rad/s" means to reduce the amplitudes of all sinusoids with frequencies above 100 rad/s.

If the signal into an ADC is a cosine wave at the radian frequency of w rad/s, its equation is given by Equation 2.1.

$$x(t) = \cos(wt) \qquad \text{(Equation 2.1)}$$

The continuous time variable is t, and A is the peak value. If the signal is sampled every T seconds, its value at the sample times is given by Equation 2.2, where n is an integer.

$$x(nT) = \cos(wnT) \qquad \text{(Equation 2.2)}$$

This sampling process is illustrated in Figure 2.1, where $T = 0.1$ and w is 2π rad/s. Column 2 of Table 2.1 gives nT, which is the sample time for every integer n, since the samples occur at $t = 0, T, 2T, \ldots$ only. A few sample values are computed in column 3 and can easily be checked using a calculator. All that has been done is to substitute nT for t, as shown in Equation 2.2. This is a valid way to get the equation of any signal after sampling, not just that of the cosine signal used here. For example, the following equations are the input and sampled output signals of an ADC for a decaying sinusoidal signal.

$$x(t) = Ae^{-3t}\cos(7t)$$

$$x(n) = x(nT) = Ae^{-3nT}\cos(7nT)$$

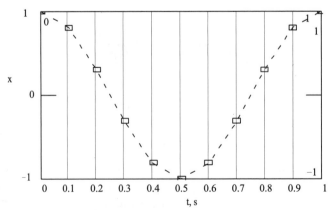

Figure 2.1. ADC samples at $nT = n(0.1)$ for slow cosine input

Effect of Signal Sampling

Table 2.1
Showing the effects of sampling the signals x(t), x1(t), and x2(t)

n	t = nT	x(n)	x1(n)	x2(n)
0	0	1	1	1
1	0.1	0.809	0.809	0.809
2	0.2	0.309	0.309	0.309
3	0.3	-0.309	-0.309	-0.309
4	0.4	-0.809	-0.809	-0.809
5	0.5	-1	-1	-1
6	0.6	-0.809	-0.809	-0.809
7	0.7	-0.309	-0.309	-0.309
8	0.8	0.309	0.309	0.309
9	0.9	0.809	0.809	0.809
10	1	1	1	1

Notice that for notational convenience the sampled signal argument is usually written without the sample period T, but T or its value is never left out of the sampled signal equation (or it would be a different equation).

2.2 Periodicity of Any DSP System Frequency Response

Let's look at the values of the sampled sinusoid when its frequency w is increased by the sampling frequency w_s, as shown in the following equation.

$$x1(t) = \cos[(w + w_s)t] \qquad \text{(Equation 2.3)}$$

The sampling frequency in Hz is $1/T$ and in rad/s is $2\pi/T$, so Equation 2.3 gives

$$x1(n) = \cos[(w + 2\pi/T)nT]$$

$$= \cos(wnT + 2n\pi)$$

$$= \cos(wnT)$$

Notice that after sampling, the signal in Equation 2.3 looks just like the original sampled signal in Equation 2.2. You can see that the sampled values of Equation 2.3 shown in column 4 of Table 2.1 are the same as the values in column 3. The preceding equations make use of the fact that the sine or cosine of an angle offset by 2π is not changed. You can easily see that any angles offset by multiples of 2π are the same by turning around 2π radians in a room and finding that you are facing in the original direction again.

If the cosine signal frequency is increased by integer multiples of the sampling frequency, its sampled values out of the ADC are indistinguishable from the corresponding values of the samples of the unshifted cosine input to the ADC. This is true even if the original frequency is decreased by multiples of the sampling frequency such that a negative argument for the sampled cosine is obtained, since $\cos(-a)$ is $\cos(a)$, from trigonometry. This is also apparent to all students in electronics, since on a scope the only difference between a cosine signal and a negative cosine signal on an oscilloscope is when the trace starts.

Why was it useful to point out that two different cosine signals into an ADC have the same sample values out of the ADC if they differ in frequency by the sampling frequency of $2\pi/T$ rad/s? The reason is that all signals worked on by a DSP system are represented in a computer as outputs from an ADC. If two signals have the same values out of the ADC, then any DSP system will do exactly the same thing to both signals. One example is a lowpass digital filter. The purpose of a lowpass filter is to reduce the amplitude of sinusoidal signals above a specified frequency, while not reducing the amplitude below a specified frequency. From the preceding discussion, we know that above a certain frequency the digital lowpass filter will start acting like a highpass filter because a high frequency sinusoidal will have the same sample values as a lower frequency sinusoidal signal. This important fact will be used in Chapter 3 in drawing graphical digital filter specifications.

The preceding equations show that all ADC outputs look alike if separated by $2\pi/T$ radians per second. There is no way around this. Whatever a digital filter does, its characteristics repeat themselves every $2\pi/T$ radians per second, because after going through an ADC the inputs look the same, as Table 2.1 shows. The math is just an explicit way to show this, but it can be seen in Figure 2.2, where the same sinusoid shown in Figure 2.1 is sampled at the original rate after its frequency is increased by $2\pi/T$ radians per second. The cosine wave is shown by dashed lines,

Effect of Signal Sampling

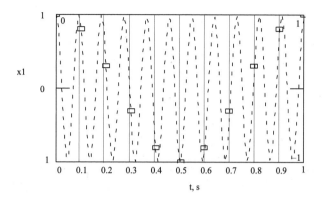

Figure 2.2. ADC samples at $nT = n(0.1)$ for fast cosine input

and the sample values by boxes. If you want to filter an analog signal, you must sample at a high enough rate or eliminate the high frequency content of the signal first.

We have shown that a cosine signal increased in frequency by the ADC sampling rate has the same sample values as if its frequency were not increased. Next it will be shown that its sample values will be the same at an even lower increased frequency. As stated earlier, from trigonometry we have $\cos(a) = \cos(-a)$. Again using the property that any trigonometric function is identical if its angle is changed by 2π radians, we will show that the signal $x2(t)$ in Equation 2.4 has the same value out of the ADC as the original signal $x(t)$.

$$x2(t) = \cos[(-w + w_s)t] \qquad \text{(Equation 2.4)}$$

This cosine signal $x2(t)$ is just the original signal $x(t)$ with its frequency at the sample frequency minus the original signal's frequency. When samples of $x2(t)$ are taken every T seconds by the ADC, the equation for the sample values is derived as in the following set of equations.

$$x2(n) = \cos[(-w + 2\pi/T)nT]$$

$$= \cos(-wnT + 2\pi n)$$

$$= \cos(-wnT)$$

$$= \cos(wnT)$$

Digital Signal Processing

Again it is seen that at a higher frequency even less than the sample frequency the sampled values out of an ADC and into a computer look identical to those at a lower frequency. This can be verified by computing the values in column 5 of Table 2.1 at the sample times on a calculator, using Equation 2.4 with $T = 0.1$ and $w = 2\pi$ rad/s. This new higher frequency is not the original increased by the sampling frequency, but the original frequency subtracted from the sample frequency. This result is true for all integer multiples of the sample frequency.

Thus, using the simple algebraic substitution of nT for t to obtain the value of a cosine signal at the sample times separated by T seconds, we have seen that all DSP systems must do the same thing to sinusoids of frequency w as they do to sinusoids at frequencies w above and below the sample frequency, since their sample values into the computer are the same. This is shown in Figure 2.3, where a cosine of amplitude A is plotted at w, $w_s + w$, and $w_s - w$. Remember that w_s is just the sampling frequency in rad/s. This significant result must be taken into account when designing a DSP system. If the DSP system is the previously mentioned lowpass filter, its filtering characteristics repeat, as shown in Figure 2.4. When specifying the frequency characteristics of a DSP system such as a digital filter, you must be aware that they will repeat above half the sample rate at π/T in rad/s, as seen in Figure 2.4, and this repetition is periodic.

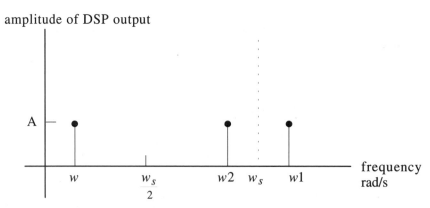

Figure 2.3. Equivalence of DSP output for sinusoids at w, w1, w2

Effect of Signal Sampling

2.3 Aliasing and Nyquist Limit

The condition where the highest input signal frequency content is equal to half the sample rate is called the **Nyquist limit**, and it leads to the **Nyquist criterion**. The Nyquist criterion is not violated if the sampling rate is more than twice as high as the highest frequency of the sinusoids in the signal, that is, if the highest input signal frequency content is less than the Nyquist limit. This is shown in Figure 2.4. The figure is drawn for an arbitrary input signal frequency spectrum; the signal might be composed of only one cosine wave or many cosine or sine waves.

In Figure 2.4 the Nyquist criterion is not violated, but it can be seen that if the sampling frequency is not greater than twice the highest frequency of any sinusoid in the signal, frequency components in the original signal would look like lower frequency signal components. Figure 2.5 shows the same signal spectrum sampled at a lower rate, so that the Nyquist limit is violated. The DSP system not only treats sinusoids above the Nyquist limit as if they were lower frequency sinusoids, but actually includes them with the actual lower frequency sinusoids. It is important to be aware of this double whammy. The DSP system not only has a periodic frequency spectrum for its output signal, but it also modifies the spectrum you have tried to design by including higher frequency sinusoids for processing as if they were the corresponding lower frequency sinusoids. There is no way to undo this effect of a DSP system. You must either be aware of the damage and accept the consequences, avoid violating the Nyquist criterion by sampling faster, or eliminate frequency components in the signal above the Nyquist limit.

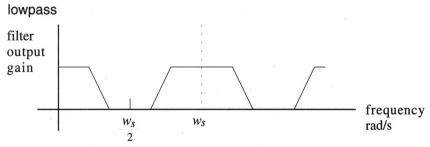

Figure 2.4 Lowpass digital filter magnitude spectrum

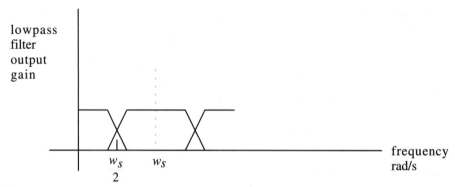

Figure 2.5. Lowpass digital filter showing Nyquist violated

2.4 Anti-aliasing Filters

Because the frequency content of most signals is unknown to some degree, especially when noise is considered, most DSP systems use a lowpass analog filter called an **anti-aliasing filter** in front of the ADC. This filter must be an analog filter, since it is in front of the ADC. If it were after the ADC it would itself be a digital filter, with the same problems you want to eliminate from the original DSP system! Figure 2.6 shows a typical DSP system using an anti-aliasing filter.

The specifications on the anti-aliasing filter depend on the input signal sinusoidal frequency content and the proposed sampling rate specified by the sample period T. As mentioned earlier, it must be an analog lowpass filter. It seems strange that almost all DSP systems and especially digital filters include an analog filter. However, this is usually a very simple lowpass filter to build. All it needs to do is to reduce the amplitudes of sinusoids in the signal into the ADC below an acceptable level above a frequency at which they look like significant lower frequency sinusoids.

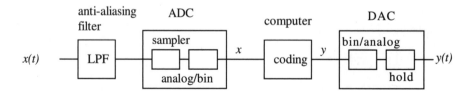

Figure 2.6. DSP system with anti-aliasing filter

Effect of Signal Sampling

Figure 2.7 shows this process for a digital lowpass filter, with the much less stringent anti-aliasing lowpass analog filter response shown in dashed lines.

Example 2.1. Determining the requirements for an anti-aliasing filter

Problem: Assume that a digital lowpass filter is to be designed to pass all frequencies below 100 rad/s and reduce all frequencies above 500 rad/s by 32. Let's assume the sample period is 0.001 s. An anti-aliasing filter must be designed for this digital filter.

Solution: The sample rate in rad/s is 2000π. From Figure 2.7 it can be seen that frequencies above 2000π minus 100 rad/s must be reduced by 32 (by the anti-aliasing filter) or else the digital filter will pass them as if they were the corresponding low frequency signals in the passband.

The requirements on the anti-aliasing filter are seen to be that it reduces the amplitude of the signal into the ADC by 32 or more above 5783 rad/s while not significantly reducing the frequencies below 100 rad/s. From an analog filtering course, it can be learned that this is easily achieved by a first-order lowpass analog filter. A first-order filter reduces the signal by 2 every time the frequency doubles beyond the corner frequency. If the corner frequency is set at 100 rad/s, by the time the frequency is 3200 rad/s (which is five doublings of the corner frequency), the analog signal is reduced by 32. This first-order lowpass filter could even be a simple two-component RC filter.

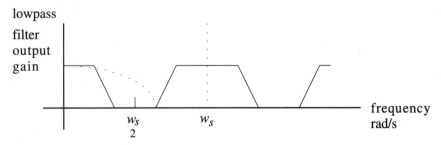

Figure 2.7 Lowpass digital filter and anti-aliasing filter

Digital Signal Processing

2.5 The Nyquist Limit and DSP Output Periodicity by Mathematical Means

Most DSP texts and most engineers have been trained to think of aliasing and the Nyquist criterion in terms of the frequency of a sampled signal. A simple derivation is given in the following discussion; most other derivations are much more complex. It is not necessary to determine the magnitude spectrum of a sampled signal, since we have determined aliasing and the Nyquist criterion without it. Inside the computer or DSP chip, there are just a bunch of numbers being manipulated, not a signal with a sampled analog spectrum. But to illustrate the other approach, the following derivation is given, using knowledge from first-semester calculus and Laplace transform theory. The output of an ADC is a sampled signal and we will show that this leads to it having a periodic spectrum with a period of $1/T$ Hz or $2\pi/T$ rad/s. Again, as can be seen in Figure 2.4 and Figure 2.5, this leads to the Nyquist criterion on the sampling rate.

We will use the delta or impulse function $\delta(t)$ used in analog signal processing class, a very narrow and tall signal with area (or strength) = 1 at $t = 0$ and zero value anywhere else. Then $\delta(t - T)$ is just a spike of strength 1 at $t = T$ and zero everywhere else. The sum of $\delta(t)$ and $\delta(t - T)$ is just spikes of strength 1 at $t = 0$ and another at $t = T$. Using this approach, the output of an ADC is given by the following equations.

$$x(nT) = x(n) = x(0)\delta(t) + x(T)\delta(t - T) + x(2T)\delta(t - 2T) + \ldots$$

$$x(n) = \sum_{n=0}^{\infty} x(nT)\delta(t - nT) = x(t)\sum_{n=0}^{\infty} \delta(t - nT)$$

Now it can be seen that $\sum_{n=0}^{\infty} \delta(t - nT)$ is periodic (plot it), so it has a Fourier series, as given by the following equation.

$$\sum_{n=0}^{\infty} \delta(t - nT) = a_0 + \sum_{k=1}^{\infty} a_k \cos(k\omega_s t) + \sum_{k=1}^{\infty} b_k \sin(k\omega_s t), \text{ where } \omega_s = 2\pi/T$$

The coefficients for the preceding Fourier series are computed as follows, using the standard formulas for computing Fourier series coefficients.

Effect of Signal Sampling

$$a_0 = \frac{1}{T}\int_{-T/2}^{T/2} \sum_{n=0}^{\infty} \delta(t - nT)\,dt = 1/T$$

$$a_k = \frac{2}{T}\int_{-T/2}^{T/2} \sum_{n=0}^{\infty} \delta(t - nT) \cos(k\omega_s t)\,dt = \frac{2}{T}$$

$$b_k = \frac{2}{T}\int_{-T/2}^{T/2} \sum_{n=0}^{\infty} \delta(t - nT) \sin(k\omega_s t)\,dt = 0$$

Using the preceding coefficients in the Fourier series equation, we can write the equation for the sum of the impulses in the following form.

$$\sum_{n=0}^{\infty} \delta(t - nT) = \frac{1}{T} + \frac{2}{T}\sum_{k=1}^{\infty} \cos(k\omega_s t)$$

Using this in the equation for $x(nT) = x(n)$, as a discrete-time signal, gives the following equation.

$$x(n) = x(t)[\frac{1}{T} + \frac{2}{T}\sum_{k=1}^{\infty} \cos(k\omega_s t)]$$

Now let $x(t)$ be any sinusoid of amplitude A at frequency w. Then the preceding equation for $x(n)$ can be written in the following form.

$$x(n) = x(nT) = A\cos(\omega t)[\frac{1}{T} + \frac{2}{T}\sum_{k=1}^{\infty} \cos(k\omega_s t)]$$

Using the trigonometric identity for the product of cosines, we finally have the result we need in the following equation.

$$x(n) = x(nT) = \frac{A}{T}\cos(\omega t) + \frac{2A}{T}\sum_{k=1}^{\infty} [0.5\cos(\omega - k\omega_s)t + 0.5\cos(\omega + k\omega_s)t]$$

From the preceding equation it is obvious that the sampled signal frequency response as a discrete time signal is periodic and symmetric every w_s rad/s or $1/T$ Hz and looks like Figure 2.3 for any w, where $w1$ is $w + w_s$ and $w2$ is $-w + w_s$.

Summary

In this chapter the effect of periodic sampling of an analog signal is shown to generate two important characteristics. These characteristics were developed by using a test signal composed of only a cosine wave, but any signal can be considered to be composed of a sum of sinusoids like the cosine wave. This is known from Fourier analysis and is also obvious to students of filtering, since otherwise there would be no need to design filters to amplify or reduce certain frequencies.

The first important effect of sampling is that any sinusoid that is sampled has the same sample values as a sinusoid offset in frequency by the original signal frequency above and below the sample frequency. Thus the output of any DSP system must be periodic about the sample frequency, and also any multiple of the sample frequency. This is due to the fact that if the sample values are the same when input into the DSP chip or computer, it will do the same thing to them.

The other effect of sampling an analog signal is a result of the repetition of the DSP frequency characteristics just discussed and leads to the Nyquist criterion. If a higher frequency sinusoid is treated in the DSP system the same as a lower frequency sinusoid, then the DSP system output will be the result of both sinusoids, but it should be the result of only one. To avoid this effect, the maximum frequency of a sinusoid should be less than half the sampling frequency, which is called the Nyquist limit. If the input signal has frequencies above this limit, it is said that the Nyquist criterion is violated. If this is the case, then an analog anti-aliasing filter is used to eliminate the higher frequency sinusoids that look like lower frequency sinusoids after sampling, so that the Nyquist criterion is not violated. This anti-aliasing filter is a simple lowpass analog filter.

In developing the effects of sampling a signal into an ADC, we showed that the analog signal in could easily be modified to give the equation of the ADC output signal. All that needs to be done is to replace the time variable t by nT. Also, the T in nT is dropped when it is in the argument

Effect of Signal Sampling

of a signal for notational convenience, but it is never dropped in any other place.

Self-Test

1. Change the equations for the following signals to describe the signals after they go through an ADC with a sample period of T seconds.

   (a) $x(t) = e^{-3t}$

   (b) $x(t) = 5t^2$

2. Compute the value of the sample for $n = 10$ for the following signals after they have gone through an ADC with the sample time $T = 0.05$ seconds.

   (a) $x(t) = 7\sin(25t)$

   (b) $x(t) = 2\cos(50t) - 4\cos(100t)$

3. Compute the values of the following signals after going through an ADC with $T = 0.1$ s for the values of n from 0 to 10.

   (a) $x(t) = 2\cos(10t)$

   (b) $x(t) = 2\cos(72.83t)$

4. For a digital filter system with the given ADC sample periods T, compute the Nyquist limit.

   (a) $T = 0.1$ s

   (b) $T = 0.002$ s

5. Determine which input signals to a digital filter or DSP system will be aliased by the given sample period T.

   (a) $x(t) = 2\cos(10t)$, $T = 0.1$ s

   (b) $x(t) = 8\cos(15t)$, $T = 0.2$ s

6. Determine whether the following signals will be aliased for the given sample period. If the signal is aliased into having the same sample values as a lower frequency sinusoidal signal, determine that lower sinusoidal signal.

   (a) $x(t) = 7\cos(25t)$, $T = 0.1$ s

   (b) $x(t) = 3\sin(37t)$, $T = 0.15$ s

   (c) $x(t) = 5\cos(160t)$, $T = 0.02$ s

7. Determine the equation $x(n)$ for the following signal $x(t)$, using only one cosine term, after it is sampled with a sample period of $T = 0.1$ s. Hint: The higher frequency sinusoid is aliased to what?

   $x(t) = 3\cos(7t) + 3\cos(69.83t)$

Problems

1. Change the equations for the following signals to describe the signals after they go through an ADC with a sample period of T seconds.

   (a) $x(t) = 3e^{-7t}$

   (b) $x(t) = 5\sin(3t)$

2. Compute the value of the sample for $n = 6$ for the following signals after they have gone through an ADC with the sample period $T = 0.02$ seconds.

   (a) $x(t) = 12\cos(3t)$

   (b) $x(t) = 7 - 8e^{-2t}$

3. Compute the values of the following signals after going through an ADC with $T = 0.05$ s for the values of n from 0 to 3.

   (a) $x(t) = 0.25t^2$

   (b) $x(t) = 3\sin(20t) - 5\cos(40t)$

4. For a digital filter system with the given ADC sample periods T, compute the Nyquist limit.

   (a) $T = 0.025$ s

   (b) $T = .001$ s

5. Determine which input signals to a digital filter or DSP system will be aliased by the given sample period T.

   (a) $x(t) = -2\cos(10t)$, $T = 0.3$ s

   (b) $x(t) = 4\sin(105t)$, $T = 0.03$ s

6. Determine whether the following signals will be aliased for the given sample period. If the signal is aliased into having the same sample values as a lower frequency sinusoidal signal, determine that lower sinusoidal signal.

   (a) $x(t) = 17\sin(25t)$, $T = 0.1$ s

   (b) $x(t) = 4\cos(3t) + 2.5\sin(100t)$, $T = 0.05$ s

   (c) $x(t) = 5\cos(160t)$, $T = 0.02$ s

7. Determine the equation $x(n)$ for the following signal $x(t)$, using only one cosine term after it is sampled with a sample period of $T = 0.003$ s. Hint: The higher frequency sinusoid is aliased to what?

   $x(t) = 2\cos(15t) + 2\cos(2079.4t)$

Answers to Self-Test

1a. $x(n) = e^{-3nT}$

1b. $x(n) = 5(nT)^2$

2a. $x(10) = -0.464$

2b. $x(10) = -1.877$

3. $x(0) = 2.0$, $x(1) = 1.08$

4a. 31.4 rad/s

4b. 1570.7 rad/s

5a. not aliased

5b. aliased

6a. not aliased

6b. aliased, $-3\sin(4.89t)$

6c. aliased, $5\cos(154.2t)$

7. $x(n) = 6\cos(0.7n)$

chapter 3

Digital Filter Specifications

Introduction

In this chapter we begin the first step in designing digital filters, which is drawing their graphical specifications. From these specifications we will later learn how to determine the a and b coefficients for the digital filter. A digital filter graphical specification is just an ideal plot versus the frequency of where the gain curve of the digital filter is allowed to go and not allowed to go. We need to define gain, which is the single most important characteristic of any filter. We will also define and use the usual axes of the gain plot of filters, as well as define the gain in dB.

There are four basic types of filter graphical specifications, one for each of the four basic filter types: lowpass, highpass, bandpass, and bandstop. Because of the periodicity of digital filters, a bandpass digital filter gain plot may look like that of a highpass digital filter. Similarly, a lowpass digital filter may have the identical gain plot of a bandstop digital filter. The only way to distinguish them is to know the sampling frequency. Knowledge of the sampling frequency used to specify each filter graphical specification is essential. Two different specifications may look alike, but they will behave differently because of aliasing!

3.1   Introduction to Filter Gain, Loss, dB, and Graphical Filter Specifications

The **gain** is defined below. Because it was shown in Chapter 2 that all DSP systems start to repeat their output spectrums in magnitude every 0.5 times the sampling frequency, our discussion will usually limit the frequency axis to this value. But remember that many texts and digital filter programs do not, leaving it up to the user to interpret the plot based on the sampling frequency. Also, we will be concerned only with the ratio of the amplitude of the output with respect to the input amplitude for sinusoidal inputs at the same frequency. This is the magnitude transfer function or gain, as shown in the following equation.

$$\text{gain} = \text{magnitude transfer function} = \frac{\text{amplitude of output sinusoid at frequency } w}{\text{amplitude of input sinusoid at frequency } w}$$

This simple definition of filter gain illustrates important ideas about filters. The first is that digital filter designers, as opposed to digital control designers, are usually interested only in sinusoidal signals. The second is that digital filter designers are usually interested only in the amplitudes of sinusoidal signals and not in the phase (the phase is extremely significant for digital control). Finally, the gain is a ratio of output amplitude over input amplitude for the same frequency. With a plot of gain versus frequency of a filter, it is easy to see what the filter does to the output by multiplying the input amplitude at any frequency by the gain at that frequency. As we will soon see, the gain is the vertical axis of the graphical specification.

In almost all engineering and especially in communications, gain is specified in **dB**, which is defined in the following equation. This terminology allows a wider range of gain values to be plotted on a graph.

$$\text{gain}_{dB} = 20 \log(\text{gain})$$

Thus if the output amplitude is 10 times the input amplitude at the same frequency, the gain is 10 and the gain_{dB} is 20. Every time the gain is 10 times what it was, the gain in dB increases by 20 dB. Note that since gain in dB uses logarithms, a gain of 0 has no equivalent in gain_{dB}. If gain in dB is positive, it means that the gain is greater than 1; and if gain in dB is negative it means that the gain is less than 1. These statements are illustrated in Example 3.1.

Digital Filter Specifications

Example 3.1. Computing gain$_{dB}$ from gain

Problem: Let's assume that the gain of a filter at each of the following frequencies is to be converted to gain in dB.

frequency, rad/s	gain
0.1	2
0.5	1
1	0.5
5	0.1
20	1
40	10
100	100

Solution: Taking the logarithm of each gain and multiplying it by 20 gives the results shown in the following columns.

frequency, rad/s	gain$_{dB}$
0.1	6
0.5	0
1	−6
5	−20
20	0
40	20
100	40

It is usual in engineering to plot the frequency along the horizontal axis, using a log scale, since most filter properties are specified in ratios of the frequency, such as dropping 20 dB per decade. This means the gain reduces by a factor of 10 every time the frequency increases by a factor of 10. By then plotting gain in dB and frequency as if on semi-log paper, every time any distance on either axis is doubled, that value is multiplied by 10. This allows a wider range of frequency and gain values to be plotted. Note again, however, that by plotting the gain in dB, a negative value in dB means the gain was less than 1. Using a log base 10 scale on the frequency axis means there will never be a zero frequency. Usually there is no need to compute the log of the frequency, since most programs print out the frequency axis scaled to logarithms already.

Sometimes the magnitude of the transfer function is specified as **loss**, which is just the inverse of the gain. In terms of dB, loss is just the negative of the gain in dB, as shown in the following equation.

$$\text{loss} = (\text{gain})^{-1} \text{ and } \text{loss}_{dB} = -\text{gain}_{dB}$$

3.2   The Lowpass Digital Filter Specification

One of the easiest graphical specifications to draw is that for the **lowpass** digital filter. From analog filtering the student remembers that a lowpass filter is supposed to pass (or not reduce very much) low frequency signals, while it should stop (or reduce greatly) frequencies above a specified frequency. The band of frequencies specified to be passed by the filter is called the **passband**, while the range of frequencies specified to be stopped by the filter is called the **stopband**. As saying this is cumbersome and loaded with ambiguities, almost always a graphical specification is drawn, using the following definitions:

g_{pmax} is the maximum allowed gain in the passband.

g_{pmin} is the minimum allowed gain in the passband.

g_{smax} is the maximum allowed gain in the stopband.

w_p = the highest frequency in the passband

w_s = the lowest frequency in the stopband

w_f = the folding frequency or half the sampling frequency = π/T in rad/s

Using these definitions, the graphical specification for a lowpass digital filter looks like that in Figure 3.1, where also the frequency axis stops at $w = w_f$, since all DSP system gains will repeat after that, whether you like it or not.

In Figure 3.1 the forbidden regions are shown as shaded blocks. Any filter with a gain versus frequency within the clear regions is a lowpass digital filter that satisfies the specifications. There is no minimum stopband gain since any gain below the maximum is even better. The region between the passband and the stopband is called the **transition band**. The narrower it is, the better the filter is usually considered to be. However, in succeeding chapters we will see that just as in the analog filter case, the narrower the transition band, the more complex the filter design.

Digital Filter Specifications

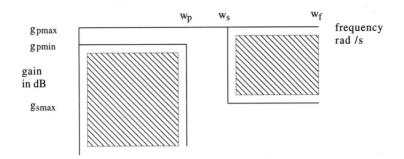

Figure 3-1. The general lowpass filter graphical specification

Example 3.2. Specification of a digital lowpass filter

Problem: Let's assume that the customer wants a digital filter that will not amplify the signal at all in the passband while not reducing the gain by more than 3 dB in the passband. The passband extends out to 1,000 rad/s and the stopband starts at 10,000 rad/s. The digital filter is to reduce the gain in the stopband by at least 40 dB. The sampling frequency is given as 40,000 rad/s, or T is 0.16 ms.

Solution: Given these specifications, the graphical specification for the digital lowpass filter is shown in Figure 3.2, with the frequency axis only going out to 20,000 rad/s, since any filter gain will repeat after that.

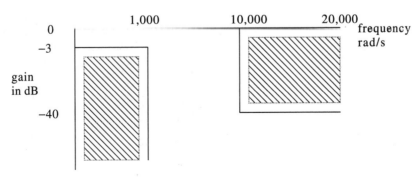

Figure 3.2. Lowpass filter graphical specification of Example 3.2

Digital Signal Processing

3.3 The Highpass Digital Filter Specification

The **highpass** filter graphical specification is also not too difficult to draw. A highpass filter is a filter that stops (or greatly reduces) frequencies below a specified frequency, but passes (or changes very little) the frequencies above a specified frequency. The graphical specification for a highpass digital filter can be drawn using the preceding definitions for the lowpass digital filter specification, except that w_p is the lowest frequency in the passband and w_s is the highest frequency in the stopband. This is shown, in general, in Figure 3.3, remembering that there is no need to extend the specification beyond half the sample frequency of w_f.

Example 3.3. Specification of a highpass digital filter

Problem: Let's assume the customer wants a digital filter running (reads the ADC and outputs to the DAC) at 500 Hz. This is $1{,}000\pi$ rad/s. The filter is required to reduce input frequencies below 200 rad/s by more than 20 dB, but not reduce any input frequencies above 500 rad/s by more than 1 dB and not increase any signals above that.

Solution: The graphical specification for this filter is shown in Figure 3.4. Any filter designed using the methods shown in Chapters 7, 10, and 11 that has a gain in the clear area meets the customer's specification for the filter.

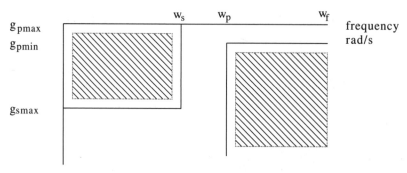

Figure 3.3. The general highpass filter graphical specification

Digital Filter Specifications

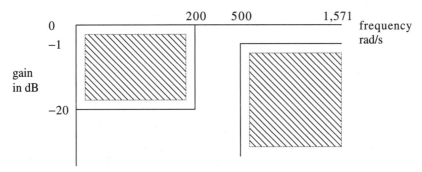

Figure 3.4. The highpass filter specification for Example 3.3

3.4 The Bandpass Digital Filter Specification

The digital bandpass filter specification is a little more complex than the previous two graphical specifications simply because the allowed filter gain region is more complex. A **bandpass** filter is one that passes (or changes very little) frequencies that are between two specified frequencies, while it stops (or greatly reduces) frequencies above and below two other specified frequencies. In order to draw the graphical specifications for a bandpass filter, the following definitions are needed.

g_{smax} is the maximum allowed gain in the stopbands.

g_{pmax} is the maximum allowed gain in the passband.

g_{pmin} is the minimum allowed gain in the passband.

w_{s1} is the upper frequency limit of the lower stopband.

w_{p1} is the lower frequency limit of the passband.

w_{p2} is the upper frequency limit of the passband.

w_{s2} is the lower frequency limit of the upper stopband.

w_f is half the sampling frequency.

Digital Signal Processing

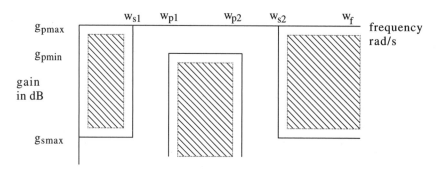

Figure 3.5. The general bandpass graphical specification

Although this type of filter specification could be made even more complex by specifying different maximum gains for each stop band, this is not usually done. Notice that now there are two stopbands and two transition bands. The general graphical specification for a bandpass digital filter is shown in Figure 3.5. Example 3.4 shows how to draw the bandpass graphical specification.

Example 3.4. Specification of a bandpass digital filter

Problem: The customer's requirements are to design a digital filter with the time between samples equal to 0.0005 s. The filter is to reduce all frequencies below 10 Hz and above 500 Hz by more than 60 dB, while not reducing the frequencies between 50 Hz and 100 Hz by more than 2 dB. The filter should also not increase any frequency in the passband by more than 1 dB.

Solution: First the sample time T is used to determine the folding frequency.

$$w_f = \frac{\pi}{T} = 6283 \text{ rad/s}$$

All the other frequencies are multiplied by 2π to convert to rad/s. Then the graphical specification for the filter is drawn, as in Figure 3.6.

Digital Filter Specifications

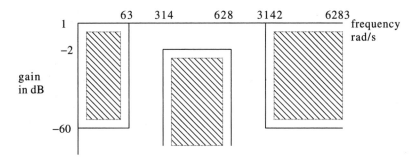

Figure 3.6. The bandpass specification for Example 3.4

From Figures 3.3 or 3.4, it can be seen that the graphical specification for a highpass digital filter that is plotted out to the sampling frequency would have the same form as a bandpass filter graphical specification plotted only out to half the sampling frequency. In fact, if the bandpass filter were perfectly symmetrical in frequency (not log of the frequency), it would look exactly like a highpass filter plotted out to its sampling frequency. The only way to be sure is to know what the sampling frequency is and where repetition begins.

3.5   The Bandstop Digital Filter Specification

The **bandstop** filter is used to stop (or greatly reduce) all frequencies between two specified frequencies, while passing (or reducing very little) frequencies below a specified frequency and above another specified frequency. The definitions given for the bandpass filter can be used for the stopband filter, since the only difference is that now the two stopband frequencies are between the two passband frequencies. Notice that now there is one stopband, two passbands, and two transition bands. Figure 3.7 shows the general stopband filter graphical specification. Again, any filter that we learn to design in later chapters with a gain in the clear region will meet the specification for the filter.

Example 3.5. Specification of a digital bandstop filter

Problem: The requirements of the filter are that the input signal is sampled at 10,000 rad/s, the frequencies in the input between 1,000

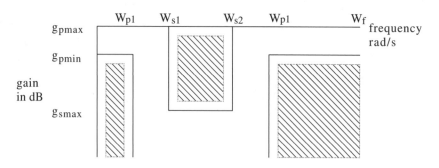

Figure 3.7. The general bandstop filter specification

and 2,000 rad/s are to be reduced by at least 40 dB, and the filter is not to increase or decrease frequencies below 500 or above 4,000 rad/s by more than 2 dB.

Solution: The graphical specification for this bandstop digital filter is drawn in Figure 3.8.

The same comment that was made about the relationship of the graphical specifications of highpass and bandpass filters also applies to the relationship between bandstop and lowpass digital filter graphical specifications. Because digital filters repeat beyond half the sampling frequency, the graphical specifications for a digital lowpass filter plotted out to its sampling frequency would look like the specifications for a symmetrical

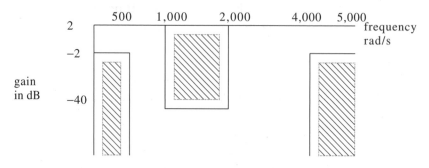

Figure 3.8. Bandstop filter specification for Example 3.5

Digital Filter Specifications

bandstop digital filter. The sampling frequency tells you if the drawing is a repetition of the lowpass filter specification or a bandstop filter.

3.6 Alternate Graphical Specifications

Many digital graphical specifications are drawn with the horizontal axis representing the frequency multiplied by the sample period rather than just the frequency of the gain function. This is just a scaling so that all filters that actually do the same thing with respect to the sampling frequency will look the same, since multiplying by T is dividing by $1/f_s$, the sampling frequency in Hz. This new frequency is called the **digital** or **scaled frequency**. It is a fictitious frequency, used only for convenience to specify stop and pass frequencies in terms of fractions of the sampling frequency.

This scaling is helpful in determining the filter coefficients in later chapters. To see what a graphical specification says about the gain at a real frequency, you only need to divide the graph frequency by T. Since the real frequency into or out of a digital filter is multiplied by T, the maximum frequency before the gain repeats at w_f is π, as shown in the following equation.

$$w_f(\text{scaled}) = w_f(\text{rad/s}) * T = 0.5 w_s T = 0.5(\frac{2\pi}{T})T = \pi$$

Thus all graphical specifications when scaled by T go from 0 to π rad/s. The pass and stop frequency specifications also need to be given in terms of the scaled frequency instead of in rad/s. This amounts to multiplying the original stop and pass frequency specifications by T also. Example 3.6 is just Example 3.5 replotted with the frequency axis scaled by multiplying by T.

Example 3.6. Graphical specification of a digital stopband filter using scaled frequency

Problem: The desired digital filter graphical specifications are given by Example 3.5, but the graphical specification is to be drawn using scaled frequencies.

Solution: Since the sampling rate in Example 3.5 was 10,000 rad/s, which is $2\pi/T$, we have that $T = 0.628$ ms. When all the frequency specification values are multiplied by this number, the new graphical specification is that shown in Figure 3.9, using scaled frequency values on the horizontal axis.

Summary

In Chapter 3 we learned how to draw the graphical specifications of digital filters. The vertical axis is gain, usually in dB, and the horizontal axis is frequency, which is usually scaled logarithmically. The graphical specifications for the four basic types of filters were developed and illustrated. Because digital filters repeated their gain characteristics past the folding frequency (half the sampling frequency), most digital filter plots only go out to that frequency. This does not mean they have no gain out there; it means the gain is a repetition of the lower frequency gain. Also remember that any frequency input to the digital filter above the folding frequency will be added to the corresponding frequency below the folding frequency before the filter works on it.

Finally we defined the scaled frequency, which is the actual input sinusoid frequency of interest multiplied by the sampling period T. By doing this, all graphical specifications and filter gain plots repeat above π rad/s. This is just a fictitious frequency, but it puts all filter specifications relative to the sampling frequency.

Problems

1. Draw the graphical specification of a digital lowpass filter out to the folding frequency in rad/s that will not reduce the gain of frequen-

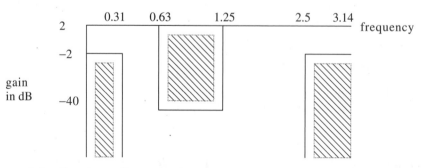

Figure 3.9. Bandstop filter specification for Example 3.6

cies below 50 rad/s by more than 2 dB, while reducing the gain of frequencies above 100 rad/s by more than 20 dB. The sampling rate is 500 rad/s.

2. Draw the graphical specification for a digital highpass filter out to the folding frequency in rad/s that will not change the gain above 500 rad/s by more than +/− 3 dB, while reducing the gain below 200 rad/s by more than 40 dB. The sampling time $T = 0.001$ s.

3. Draw the graphical specification for Problem 2, but use scaled frequencies.

4. Draw the graphical specification for a bandpass digital filter out to its folding frequency that will not reduce the gain between 100 and 200 rad/s by more than 1 dB, but will reduce the gain above 400 rad/s and below 50 rad/s by more than 25 dB. The sampling time $T = 0.0005$ s.

5. Draw the graphical specification for a stopband digital filter out to its folding frequency that will reduce the gain between 1,000 rad/s and 5,000 rad/s by more than 60 dB, but will not reduce the gain above 10,000 rad/s or below 150 rad/s by more than 3 dB. The sampling rate is 10,000 Hz.

6. Repeat Problem 5 using scaled frequencies.

7. Draw the graphical specification for a highpass digital filter out to its folding frequency in rad/s that will keep the gain above 500 rad/s between 1 and −3 dB, while reducing the gain below 100 rad/s below 35 dB. The sampling period T = 3.14 ms.

8. Draw the graphical specification for a lowpass digital filter out to its sampling frequency in rad/s that will not reduce the gain more than 4 dB below 250 rad/s, while reducing the gain above 1,000 rad/s by more than 45 dB. The sample period $T = 1.57$ ms.

9. Draw the graphical specification for Problem 1 out to 500 rad/s. If this were the graphical specification for a sampling rate of 1,000 rad/s, state the type of filter for which it is a graphical specification.

10. Draw the graphical specification for Problem 2 out to the sampling frequency. If this were now the graphical specification for a sampling

rate of 2 kHz, state the type of filter for which it is a graphical specification.

11. Draw the graphical specification for Problem 4 with the frequency axis scaled in Hz and the sampling time $T = 0.001$.

12. Draw the graphical specification for Problem 5 with the frequency axis scaled in Hz and the sampling rate is 5,000 Hz.

13. Draw the graphical specification for Problem 1 in terms of loss in dB.

14. Draw the graphical specification for problem 2 in terms of loss in dB.

chapter 4

z-Transforms

Introduction

We have shown the equation coded for a digital filter in Chapter 1, and in Chapter 2 we showed how to get the discrete or sampled time equation of a signal that is the input or output of a digital filter. However, not all the math representations have been given. In order to analyze or design a digital filter or any other DSP system, an equation of the system itself is required, not just the DSP input-output equation. We get this system equation by using the z-transforms of the sampled signals, just as analog system transfer functions are obtained from the Laplace transforms of signals. The nice thing about understanding and obtaining z-transforms is that it involves only algebra, whereas Laplace transforms involve integration from calculus.

4.1 The Need for z-Transforms of the DSP Equation

This chapter defines and shows how to obtain the z-transforms of any sampled signal. As some of the results of taking the z-transforms of specific sampled time signals are listed in Table 4.1, the z-transforms of many signals will have to be computed only once. Then, just as for Laplace transforms, a transfer function will be obtained in Chapter 5 using the ratio of the output over the input signals of the DSP system (both signals are z-transformed). This is a necessary evil to get the mathematical description of the sampled system. The transfer function must be the

Digital Signal Processing

system description, since by the preceding definition, if you multiply it by the z-transform of any input signal, you get the z-transform of the output signal as shown hereafter, with capitals representing the z-transforms of the signals.

$$Y(z) = \frac{Y(z)}{X(z)} X(z) = T(z) X(z)$$

In the equation above, $x(t)$ is the input into the ADC and $y(t)$ is the output from the DAC and $T(z)$ must be the DSP system description, since if you multiply it by the z-transform of the input, you get the z-transform of the output.

The reason for using z-transforms is that the preceding equation is valid. If you took the ratio of sampled output signal to sampled input signal to a DSP system, you would get some expression, but it would change for each input signal. By using the z-transforms of the signals, the ratio of any output to the corresponding input is always the same expression and must be the system math description, since when it is multiplied by the input, you get the output. With a math description of a digital filter, it is then possible to analyze its characteristics and even design a filter to meet the desired graphical specifications shown in Chapter 3.

4.2 The Definition of the z-Transform and Its Use

The definition of the **z-transform** of a sampled signal $f(n)$ is $F(z)$, as defined by Equation 4.1.

$$F(z) = \sum_{n=-\infty}^{\infty} f(n) z^{-n} = Z[f(n)] = \cdots + f(-1)z^{1} + f(0) + f(a)z^{-1}$$

$$+ f(2)z^{2} \cdots$$

(Equation 4.1)

The sampled signal is given by $f(n)$ as shown in Chapter 2, and $f(0), f(1)$ are the sample values at $t = 0$ and $t = 1T$, and so on. If the signal out of an ADC consists of a few samples, the z-transform of this signal, $F(z)$, is easy to get, as shown in the Example 4.1. One important property of a z-transformed signal using Equation 4.1 is that no n term will occur in the z-transformed signal. This is a good thing to check. If an equation has been z-transformed and the sample number n appears anywhere in the equation, a mistake has been made!

z-Transforms

Example 4.1. The z-transform of a signal with only a few sample values

Problem: Let the input signal $f(t)$ into an ADC be $2t$ for t greater than zero and less than 4, and zero otherwise. Let's find the z-transform of this short signal when $T = 1$.

Solution: The sample times occur at integer values of t, and the only nonzero output samples of the ADC will be $f(1) = 2$, $f(2) = 4$, $f(3) = 6$. Using the definition for the z-transform of a signal in Equation 4.1, it is seen that the z-transform of the signal is given by the following equation.

$$F(z) = 2z^{-1} + 4z^{-2} + 6z^{-3}$$

In order to determine the z-transforms of more complex signals and those that never end, we need to define two basic sampled signals and obtain their z-transforms. The first is a sampled signal that consists of just one sample, this is called the impulse or δ function. It is defined mathematically in the following equations, and as can be seen, it is not the δ function or impulse function used for continuous signals since it has a finite amplitude of 1.

$$\delta(nt) = \delta(n) = 0, \ n \neq 0$$

$$\delta(0) = 1$$

The other basic sampled signal that needs to be defined is the sampled unit step function $u(n)$, as given in the following equation.

$$u(nT) = u(n) = 1, \ n \geq 0$$

Notice again the shorthand notation of replacing the argument in each function by n, since it is understood that $u(n)$ occurs at $t = nT$.

It is easily shown in the following equations that the z-transform of the impulse function is $Z[\delta(n)] = 1$, since $\delta(0) = 1$ and all other sample values are 0.

$$Z[\delta(n)] = \cdots + \delta(-1)z^{1} + \delta(0)z^{0} + \delta(1)z^{-1} + \cdots$$

$$= 1z^{0}$$

$$= 1$$

Digital Signal Processing

Also the z-transform of an impulse function shifted by $t = kT$, where k is an integer, is easy to obtain, as shown in the following equation.

$$Z[\delta(n-k)] = \cdots + \delta(-1)z^{1-k} + \delta(n-k)z^{-k} + \delta(1)z^{-k-1} + \cdots = z^{-k}$$

The preceding result is obtained since the impulse function is zero everywhere the argument is not zero.

The z-transform of the impulse allows the discrete time equation of a few samples to be written mathematically and then the z-transform can be taken, as shown in Example 4.2, or the discrete time equation can easily be written from the z-transform.

Example 4.2. Using the impulse function for short sampled signal description

Problem: Use the impulse function to describe a sampled signal where the initial sample $x(0) = 2$, $x(3) = -2$, and the rest of the samples are zero.

Solution: Using the shifting property of time signals and the fact that a sum of two time-shifted impulse functions doesn't add at the corresponding times when they are nonzero, we get

$$x(n) = 2\delta(n) - 2\delta(n-3)$$

$$X(z) = 2 - 2z^{-2}$$

The z-transform of the unit step, $u(n)$, is more difficult to obtain in a closed form, but the procedure only involves a little algebra, which is shown in the following equations.

$$Z[u(n)] + U(z) = 1 + z^{-1} + 1z^{-2} + 1z^{-3} + \cdots$$

This summation starts at $f(0) = u(0) = 1$ and goes on forever. However, if we multiply $U(z)$ by z^{-1} we get

$$z^{-1}U(z) = 1z^{-1} + 1z^{-2} + \cdots$$

If we subtract the second equation from the first equation, we get the following equations.

$$U(z) - z^{-1}U(z) = 1$$

$$U(z)[1 - z^{-1}] = 1$$

Now dividing both sides by the term in the brackets, we get the final z-transform of $u(n)$:

$$U(z) = \frac{1}{1 - z^{-1}} = \frac{z}{z - 1}$$

As can be seen, $U(z)$, the z-transform of $u(nT) = u(n)$, is not very complicated. One of the major uses of the sampled unit step is to start and stop a signal. This is shown in one of the problems at the end of the chapter.

4.3 Derivation of the Necessary z-Transform Pairs

In the preceding section, we determined the z-transforms of two basic sampled signals, the unit impulse and the sampled unit step. In this section we expand on the pairs of sampled signals and the corresponding z-transform, using basic algebra. It is useful also to relate the sampled signal coming out of an ADC to the analog signal coming in, which is shown in column 1 of Table 4.1.

The definition of the z-transform of a sampled signal shows that if any sampled signal is A times bigger, then its z-transform is A times bigger, as shown in the following equation.

$$Z[Af(n)] = \cdots + Af(-1)z^1 + Af(0) + Af(1)z^{-1} + \cdots$$

$$= A[\cdots + f(-1)z^1 + f(0) + f(1)z^{-1} + \cdots] = A Z[f(n)]$$

Using the property just defined, we already can say

$$Z[A\delta(n)] = A$$

and

$$Z[A u(n)] = \frac{Az}{z - 1}$$

Thus we now have the z-transform of a step of A after it is sampled.

Digital Signal Processing

Table 4.1
Table of z-transforms of signals

Analog Signal	Sampled Signal	Z-transformed Signal
	$A\delta(n)$	A
$Au(t)$	$Au(n)$	$\dfrac{Az}{z-1}$
$Ae^{-at}u(t)$	$Ae^{-aTn}u(n)$	$\dfrac{Az}{z-e^{-aT}}$
	$Ac^n u(n)$	$\dfrac{Az}{z-c}$, $c = e^{-aT}$
$Atu(t)$	$AnTu(n)$	$\dfrac{ATz}{(z-1)^2}$
$A\cos(wt)u(t)$	$A\cos(wTn)u(n)$	$\dfrac{Az[z-\cos(wT)]}{z^2 - 2z\cos(wT) + 1}$
$A\sin(wt)u(t)$	$A\sin(wTn)u(n)$	$\dfrac{Az\sin(wT)}{z^2 - 2z\sin(wT) + 1}$
$Ae^{-at}\cos(wt + \alpha)u(t)$	$Ac^n\cos(wTn + \alpha)$	$\dfrac{Az[z\cos(\alpha) - c\cos(\alpha - wT)]}{z^2 - 2cz\cos(wT) + c^2}$

Another very useful and common analog signal is the exponential decaying signal that starts at $t = 0$. This signal and its sampled equation are given in the following equations.

$$f(t) = Ae^{-at} u(t)$$

$$f(nT) = f(n) = Ae^{-anT} u(n)$$

$$= A(e^{-aT})^n = Ac^n, \ c = e^{-aT} = \text{a constant}$$

z-Transforms

Now using the definition of a z-transform and the z-transform of a step of A, we can get the z-transform of a sampled exponentially decaying signal by using the following equation.

$$Z[Ac^n] = \sum_{n=-\infty}^{\infty} Ac^n u(n) z^{-n} = \sum_{n=0}^{\infty} Ac^n z^{-n} = \sum_{n=0}^{\infty} A(c^{-1}z)^{-n}$$

In the preceding section we found that

$$\sum_{n=0}^{\infty} z^{-n} = \frac{z}{z-1}$$

so that we must have

$$\sum_{n=0}^{\infty} (c^{-1}z)^{-n} = \frac{c^{-1}z}{c^{-1}z-1} = \frac{z}{z-c} = \frac{z}{z-e^{-aT}}$$

Thus the z-transform of a sampled exponential decaying signal that starts at $t = 0$ is

$$Z[Ae^{-anT}] = \frac{Az}{z-c}, \text{ where } c = e^{-aT}.$$

The z-transform just given is one of the most important in DSP. Later it will be used to determine the stability and other properties of any DSP system.

Table 4.1 lists the z-transform pairs that we have obtained so far, along with some others. The mathematical derivation of some of these is done the same way as that for the exponentially decaying sampled signal, but the final closed form solution requires the use of the Euler equation, which is not given until Chapter 6. The organization of Table 4.1 is that the analog signal is given in column 1, the sampled version of the analog signal is given in column 2, and the z-transform of the sampled signal in column 2 is given in column 3. Note that the time signals are zero before $t = 0$, which is indicated by the $u(t)$ or the $u(n)$. The z-transform of these sampled signals is not the z-transform of the product of the z-transforms of each component of the product. The z-transform was obtained by the same methods shown for the decaying exponential sampled signal starting at $t = 0$. The z-transform of the product of two signals is not the product of the z-transforms. Example 4.3 illustrates the use of Table 4.1.

Example 4.3. Using Table 4.1 to get the z-transforms of signals

Problem: Let the input signal to an ADC be given below with $T = 0.5$ s.

$$x(t) = 7e^{-3t}u(t)$$

Solution: For the preceding signal, we can see that $A = 7$, $a = 3$. Thus the sampled signal and the z-transform of the sampled signal are given in the following equations.

$$x(n) = 7e^{-1.5n}u(n)$$

$$X(z) = \frac{7z}{z - e^{-1.5}}$$

4.4 Derivation of the Major z-Transform Property Using Algebra

Using basic algebra we will use the definition of the z-transform to derive the major property of z-transforms. Later this property will allow the student to go back and forth between the DSP system math description $T(z)$ in terms of the variable z and the difference equation given in Chapter 1, which is the equation that is actually coded. This property will be developed in two different ways for the student. This property is called the **shifting property**, and it relates the z-transform of sampled signals that are time-shifted versions of each other if the time shift is in integer multiples of the sample time T. Remember that any signal or function of time that is delayed by kT is written in terms of the unshifted signal $f(nT)$ as $f(nT - kT)$, or $f(n - k)$. After the first method of derivation of this property, examples using it will be given.

If you have the representation of the z-transform of a sampled signal $f(n)$ either as $F(z)$ or as a specific function of z, it is very helpful to use this to write the time-shifted sampled signal, since the difference equation in Chapter 1 is in terms of shifted and unshifted input and output sampled signals. This representation is derived next. From the definition in Equation 4.1 we have

z-Transforms

$$Z[f(n-k)] = \sum_{n=-\infty}^{\infty} f(n-k)z^{-n}$$

If we let $u = n - k$ in the expression on the right, we have

$$Z[f(n-k)] = \sum_{n=-\infty}^{\infty} f(u)z^{-(u+k)} = \sum_{u=-\infty}^{\infty} f(u)z^{-u}z^{-k} = z^{-k}\sum_{u=-\infty}^{\infty} f(u)z^{-u} = z^{-k}F(z)$$

Where we have used $n = u + k$ and in the definition z can be replaced by u without any effect (it is a symbol used as a placeholder). The significance of the preceding equation (called the shifting property) is that if the difference equation for a digital filter is given, the use of this property on the equation gives an algebraic equation for the digital filter. Another significant property is shown in later chapters when we design a digital filter or DSP system, which is mathematically expressed in terms of the variable z, we can use the property to get the difference equation to code in a computer or a DSP chip. Let's look at a few examples of using this property on a sampled signal. In the next chapter this property is used on the DSP difference equation to obtain the math description of the DSP system.

Example 4.4. Writing the z-transform of the sampled signal f(n) delayed by 3 sample periods

Problem: Let $f(n) = f(nT)$ be an arbitrary sampled signal. We want to write the equation for this signal if it is delayed in time by $3T$.

Solution: The delayed signal is written mathematically as $f(nT - 3T)$ or $f(n-3)$, where $k = 3$. A direct application of the preceding property gives the following answer, where $F(z)$ is the z-transform of $f(n)$.

$$f(n-3) = z^{-3}F(z)$$

Example 4.5. Writing the z-transform of an equation with shifted functions

Problem: Write the z-transform of the following equation of a DSP system.

$$y(n) - 2y(n-1) = 0.5x(n-1)$$

Solution: Taking the z-transform of all the signals in the equation, we get the following equation.

$$Y(z) = 2z^{-1}Y(z) + 0.5z^{-1}X(z)$$

The significance of this property of the z-transform of a signal is its use on the equation of a DSP system, as in Example 4.5, especially the digital filter equation given in Chapter 1. Remember it is composed of current and delayed inputs and delayed outputs. This kind of equation is called a difference equation, and it is the discrete-time equivalent to the analog differential equation. In Chapter 5 we will see that this property of the z-transform will allow us to write the difference equation as an algebraic equation and then solve it, to get an algebraic equation of a digital filter by getting the ratio of the z-transform of the output over the z-transform of the input, which is the transfer function $T(z)$.

Earlier we said we would develop this shifting property in another way, which is actually more intuitive. All we need to do is to look at a few simple sampled signals and compare their z-transforms with and without delays in the sampled-time domain. First let's look at a signal $f1(n)$ that consists of a single sample at nT of 5, and then at a signal, $f2(n)$ of two samples of 2 at nT and 3 at $(n+1)T$. Let $g1(n)$ be the signal $f1(n)$ delayed by $2T$ or $g1(n) = f1(n-2)$, and $g2(n)$ be the signal $f2(n)$ delayed by $5T$ or $g2(n) = f2(n-5)$. The z-transforms of all these signals are given in the following equations, using the definition given in Equation 4.1.

$$F1(z) = 5z^{-n}$$

$$F2(z) = 2z^{-n} + 3z^{-(n+1)}$$

$$G1(z) = 5z^{-(n+2)} = 5z^{-2}z^{-n} = z^{-2}F1(z)$$

$$G2(z) = 2z^{-(n+5)} + 3z^{-(n+6)} = 2z^{-5}z^{-n} + 3z^{-5}z^{-(n+1)} = z^{-5}F2(z)$$

As can easily be seen, the $G1(z)$ and $G2(z)$ z-transforms are just the corresponding z-transforms of the undelayed transforms $F1(z)$ and $F2(z)$ each multiplied by z raised to the negative power of the delay in terms of the sample time T. Mathematically this can be written in general as

$$Z[g(nT)] = Z[f(nT - kT)] = z^{-k}Z[f(nT)]$$

z-Transforms

As you can see, this is just the equation expressing the shifting property again.

Summary

In this chapter, we have introduced the z-transform of a sampled or discrete-time signal. The definition was given, and it was seen to be an algebraic equation. This definition was used to find the z-transform of an impulse signal and a sampled unit step signal. These two z-transforms will be used later to determine two digital filter approximations to analog filters. One other z-transform of a signal was determined, the z-transform of an exponential signal. This z-transform will be used later to determine the stability of digital filters.

We also determined one of the most important properties of z-transforms, the shifting property. This property will allow us to turn the difference equation of a digital filter into an algebraic equation and then determine the mathematical description of a digital filter, called its transfer function. Also, if we have determined the mathematical description, or transfer function, of a digital filter, we can use the shifting property to write the difference equation of the filter that is actually coded.

Table 4.1 gives the input signal into an ADC in column 1, the corresponding sampled signal in column 2, and the z-transform of that signal in column 3 for several signals using the definition of the z-transform. Table 4.1 can just as easily be used in the reverse direction to give the sampled time signal in column 2, given the z-transformed signal in column 3. This process will be needed in some of the following chapters.

Self-Test

1. Determine the z-transform of the following sampled signal.

$$x(n) = 5u(n)$$

2. Determine the z-transform of the following sampled signal.

$$y(n) = 3e^{-8n}u(n)$$

3. Determine the z-transform of the following analog signal after it goes through an ADC with $T = 0.01$ s.

$$x(t) = 10e^{-2t}u(t)$$

4. Determine the z-transform of the following sampled signal.

$$y(n) = 5(0.9)^n u(n)$$

5. Determine the z-transform of the following sampled signal.

$$x(n) = 1\delta(n-1) + 2\delta(n-2) + 3\delta(n-3)$$

6. Determine the z-transform of the following sampled signal.

$$x(n) = 5\cos(1.4n)u(n)$$

7. Determine the z-transform of the following analog signal after going through an ADC with $T = 0.01$ s.

$$x(t) = 21\sin(5t)u(t)$$

8. Determine the z-transform of the following sampled signal.

$$y(n) = 7(0.8)^n \cos(0.4n + 1.57)u(n)$$

9. Determine the z-transform of the following analog signal after going through an ADC with $T = 0.03$ s.

$$x(t) = 3tu(t)$$

10. Determine the z-transform of the following sampled signal.

$$x(n) = 0.34nu(n)$$

11. Determine the z-transform for the analog signal with the following Laplace transform after going through an ADC with the sampling period $T = 0.014$ s.

$$X(s) = \frac{2}{s}$$

z-Transforms

12. Determine the z-transform for the analog signal with the following Laplace transform after going through an ADC with a sampling period $T = 0.002$ s.

$$X(s) = \frac{1.4}{s+5}$$

13. Determine the z-transform for the following sampled signal.

$$x(n) = 0.1u(n) - 0.1\delta(n) - 0.1\delta(n-1)$$

14. Determine the equation of the sampled pulse $x(n)$ described here, using the sampled unit step function.

$x(n)$ is 2 for $n = 0$ through and including $n = 5$
and zero for all other n's.

15. Determine the equation of the following sampled pulse signal.

$$x(n) = -3u(n) + 3u(n-7)$$

16. Given the only nonzero sample values of the following signal, determine the z-transform of the signal.

$$x(-1) = 2,\ x(0) = -1,\ x(2) = 1,\ x(3) = -4$$

17. Determine the z-transform of the following signal if it were sampled at $T = 0.05$ s.

$$x(t) = 3e^{-7t}\cos(25t)u(t)$$

Problems

1. Determine the z-transform of the following sampled signal.

$$x(n) = -6u(n)$$

2. Determine the z-transform of the following sampled signal.

$$y(n) = 5.7e^{-5n}u(n)$$

3. Determine the z-transform of the following analog signal after it goes through an ADC with $T = 0.02$ s.

$$x(n) = -7\sin(124t)u(t)$$

4. Determine the z-transform of the following sampled signal.

$$x(n) = 9(0.89)^n u(n)$$

5. Determine the z-transform of the following sampled signal.

$$y(n) = \delta(n) - 3\delta(n-1) + 2\delta(n-4)$$

6. Determine the z-transform of the following sampled signal.

$$x(n) = -4\sin(2.5n)u(n)$$

7. Determine the z-transform of the following analog signal after going through an ADC with $T = 0.07$ s.

$$x(n) = -5\cos(25t)u(t)$$

8. Determine the z-transform of the following sampled signal.

$$x(n) = 2(0.7)^n \cos(0.8n - 0.2)u(n)$$

9. Determine the z-transform of the following signal after going through an ADC with $T = 0.05$ s.

$$x(t) = 7tu(t)$$

10. Determine the z-transform of the following sampled signal.

$$y(n) = 1.37nu(n)$$

11. Determine the z-transform for the analog signal with the following Laplace transform after going through an ADC with the sampling period $T = 0.005$ s.

$$X(s) = \frac{-15}{s}$$

z-Transforms

12. Determine the z-transform or the analog signal with the following Laplace transform after going through an ADC with a sampling period of $T = 0.025$ s.

$$X(x) = \frac{5}{s+10}$$

13. Determine the z-transform for the following sampled signal.

$$y(n) = 2.4u(n-1) + 4\delta(n)$$

14. Determine the equation of the following sampled pulse $x(n)$, using the sampled unit step function.

   $x(n)$ is -3 for $n = -1$ through and including $n = 4$ and zero for all other n's.

15. Determine the equation of the following sampled pulse signal.

$$y(n) = 7u(n+1) - 7u(n-4)$$

16. Given the only nonzero values of the following signal, determine the z-transform of the signal.

$$x(-2) = -1, \; x(1) = 2, \; x(2) = -1$$

17. Determine the z-transform of the following signal if it were sampled at $T - 0.1$ s.

$$x(t) = -7e^{-3t}\cos(17t)u(t)$$

Answers to Self-Test

1. $\dfrac{5z}{z-1}$

2. $\dfrac{3z}{z-e^{-8}}$

3. $\dfrac{10z}{z-e^{-0.02}}$

4. $\dfrac{5z}{z-0.9}$

5. $1z^{-1} + 2z^{-2} + 3z^{-3}$

6. $\dfrac{5z[z - \cos(1.4)]}{z^2 - 2z\cos(1.4) + 1}$

7. $\dfrac{21z \sin(0.05)}{z^2 - 2\cos(0.05) + 1}$

8. $\dfrac{7z[z \cos(1.57) - 0.8 \cos(1.57 - 0.4)]}{z^2 - 1.6z \cos(0.4) + 0.64}$

9. $\dfrac{0.09z}{(z-1)^2}$

10. $\dfrac{0.34z}{(z-1)^2}$

11. $\dfrac{2z}{z-1}$

12. $\dfrac{1.4z}{z - e^{0.01}}$

13. $\dfrac{0.1z^{-1}}{z-1}$

14. $x(n) = 2u(n) - 2u(n-6)$

15. $X(z) = \dfrac{-3z}{z-1} + \dfrac{3z^{-6}}{z-1} = \dfrac{3z(z^{-7} - 1)}{z-1}$

16. $X(z) = 2z - 1 + z^{-2} - 4z^{-3}$

17. $X(z) = \dfrac{3z(1 - e^{-0.35} \cos(1.25))}{z^2 - 2e^{-0.35}z \cos(1.25) + e^{-0.7}}$

chapter 5

The z-Transform of the DSP Equation

Introduction

In this chapter we apply the z-transform to the general DSP equation given in Chapter 1. First the equation is modified to a more useful form, then the z-transform is taken of the discrete time signals on both sides of the equation. The general DSP equation in either form is an input-output difference equation and cannot be solved for the output given the input except by iteration. By taking the z-transform of the equation, an algebraic equation results that can be solved explicitly for the output given the input. The terms multiplying the input to get the output must be the equation of the filter called the transfer function, which is needed to design and analyze DSP systems and digital filters. It must be remembered that it is the general DSP equation that is coded, but it is the mathematical description of the DSP system that is used to analyze and design DSP systems.

5.1 The Transformation of the Modified DSP Equation

The Modified DSP Equation

The general DSP equation given in Chapter 1 is given again in Equation 5.1. This equation assumes that a signal sample with no subscript refers to the input or output sampled value at the current time, and a negative integer subscript refers to the corresponding number of sample periods previous to the current time. This is a convenient and easy way to write the equation for Chapter 1, but here a better and more rigorous equation must be written by making a few changes to Equation 5.1.

$$y = (b_{-1}y_{-1} + \cdots + b_{-m}y_{-m}) + (ax + a_{-1}x_{-1} + \cdots + a_{-n}x_{-n}) \quad \text{(Equation 5.1)}$$

For mathematical convenience the sampled signal subscripts will be used as arguments, with the current sample being called the nth sample, so y and x go to $y(n)$ and $x(n)$ respectively. Mathematically this is no change at all, since a subscripted variable is mathematically just a function of an integer subscript. Also x_{-1} will go to $x(n-1)$, and other signals will follow this pattern. The sample values of the signals may start at an arbitrary initial sample number 0, and the current sample values refer to the nth sample after this. If these signal arguments are multiplied by the sampling period T, we get the actual time t that the sample value corresponds to, since $t = 0T$ is the initial time and $t = nT$ is the actual time at the nth sample. This will facilitate going back and forth between the analog time t and the discrete time nT to which the nth sample corresponds.

Finally, for the coefficients it is standard to drop the negative sign on the subscripts. Thus the a coefficient goes to a_0, and the b_{-1} coefficient goes to b_1. The modified DSP equation is shown in Equation 5.2.

$$\begin{aligned} y(n) &= b_1 y(n-1) + b_2 y(n-2) + \cdots + b_M y(n-M) \\ &+ a_0 x(n) + a_1 x(n-1) + \cdots + a_N x(n-N) \end{aligned} \quad \text{(Equation 5.2)}$$

Note also that the uncapitalized integers m and n have been replaced by their capitalized letters to avoid confusion with the integer value n, which now stands for the sample number and gives the actual sample time as nT, as shown in Chapter 2. Also note that Equations 5.1 and 5.2 say exactly the same thing; just the notation has been changed. Example 5.1 illustrates this modification of the general DSP equation to the new form used in this and almost all texts.

Example 5.1. The modification of a DSP equation to the more standard form

Problem: Convert the following equation into the modified or standard form.

$$y = 3y_{-1} - 5y_{-2} + 0.5x - 0.75x_{-1} + 0.25x_{-2}$$

The z-Transform of the DSP Equation

Solution: First we will let the current sample be designated with the n subscript, as shown here.

$$y_n = 3y_{n-1} - 5y_{n-2} + 0.5x_n - 0.75x_{n-1} + 0.25x_{n-2}$$

Next let the subscripts be used as arguments of the input and output signals.

$$y(n) = 3y(n-1) - 5y(n-2) + 0.5x(n) - 0.75x(n-1) + 0.25x(n-2)$$

The preceding equation is in the more standard form, and it still says the same thing as the original equation. But now if the arguments, which represent sample numbers, are multiplied by the sample period, we have the product representing the actual sample times. Also note that in the preceding equation 3 is b_1, -0.75 is a_1, and so on.

In order to show explicitly how the subscripts of the coefficients are handled to get the modified and more standard form of the general DSP equation, we leave them as arbitrary coefficients in Example 5.2. This practice is useful in understanding the difference equations in this and other texts, as well as DSP programs that list the computed coefficients alongside the a or b coefficients they represent.

Example 5.2. Getting the standard form of the DSP equation with arbitrary coefficients

Problem: Given the following digital filter equation, modify it so that it is in standard form.

$$y = b_{-1}y_{-1} + b_{-2}y_{-2} + ax + a_{-1}x_{-1} + a_{-2}x_{-2}$$

Solution: First we let the subscript n and not 0 represent the current sample number and then use the subscripts for the signals as arguments of the signals. This gives the following equation.

$$y(n) = b_{-1}y(n-1) + b_{-2}y(n-2) + ax(n) + a_{-1}x(n-1) + a_{-2}x(n-2)$$

Finally we change the signs on the subscripts of the coefficients so that they correspond to the numbers being subtracted from n in the arguments. This gives the following equation.

$$y(n) = b_1 y(n-1) + b_2 y(n-2) + a_0 x(n) + a_1 x(n-1) + a_2 x(n-2)$$

In the modified or standard form of the DSP equation, you should have the following pattern:

- The sampled input and output signals should have no subscripts to show which sample number they correspond to. Instead, the sample number is in the argument of the signal.

- As the sample number n is used to represent the current sample, the equation should solve for $y(n)$ on the left, since the equation is used to solve for the current computer or DSP chip output given the current input and previous inputs and outputs.

- The subscripts of the coefficients multiplying the input and output samples have coefficients that are the integers subtracted from the current sample number in the signal arguments.

- The coefficient of the $y(n)$ term on the left, b_0, is always 1, since the equation could be divided by this value without changing the equation.

Taking the z-Transform of the Modified DSP Equation

The equation given by Equation 5.1 or its modified equivalent is called a **difference equation**, specifically an input-output difference equation, because it is made up of sums and differences of samples of signals. The difference equation is handy to see how to code the sampled input and output signals to get the current output signal, but it is hard to actually solve for the output mathematically in closed form. To help do this, we will use z-transforms of the signals in the difference equation.

By using the property of z-transforms given in Chapter 4, we can get an algebraic equation, instead of one involving the shifted input and output values at discrete times. The shifting property is repeated in the following equation, with $F(z)$ being the z-transform of $f(n)$.

$$Z[f(n-k)] = z^{-k} Z[f(n)] = z^{-k} F(z)$$

Equation 5.3 shows this property applied to Equation 5.2, with $Y(z)$ and $X(z)$ being the symbolic representations of the z-transforms of $y(n)$ and $x(n)$ respectively.

$$Y(z) = b_1 Y(z) z^{-1} + b_2 Y(z) z^{-2} + \cdots + b_M Y(z) z^{-M}$$
$$+ a_0 X(z) + a_1 X(z) z^{-1} + \cdots + a_N X(z) z^{-N} \quad \text{(Equation 5.3)}$$

The z-Transform of the DSP Equation

Example 5.3. Taking the z-transform of a difference equation for a DSP system

Problem: Given the following difference equation, obtain the z-transformed equation by z-transforming all the signals in the equation.

$$y(n) = 0.5y(n-1) + 0.5x(n) - 0.25x(n-1)$$

Solution: By using the shifting property of z-transform, the preceding equation becomes the following equation.

$$Y(z) = 0.5Y(z)z^{-1} + 0.5X(z) - 0.25X(z)z^{-1}$$

5.2 The Transfer Function of a Digital Filter

The reason for using z-transforms is that Equation 5.3 now can be solved for $Y(z)$, the z-transform of the output DSP signal, in terms of $X(z)$, the z-transform of the input signal using algebra. This equation can then be solved for the z-transform of the output. This will lead to the mathematical description of a digital filter or any DSP system represented by the difference equation. The following equations show the steps to solve for the z-transform of the output, with the solution given by Equation 5.4.

$$Y(z) = Y(z)[b_1 z^{-1} + b_2 z^{-2} + \cdots + b_M z^{-M}] + X(z)[a_0 + a_1 z^{-1} + \cdots + a_N z^{-N}]$$

$$Y(z)[1 - b_1 z^{-1} + b_2 z^{-2} + \cdots + b_M] = X(z)[a_0 + a_1 z^{-1} + \cdots + a_N z^{-N}]$$

$$Y(z) = \frac{a_0 + a_1 z^{-1} + \cdots + a_N z^{-N}}{1 - b_1 z^{-1} - b_2 z^{-2} - \cdots - b_M z^{-M}} X(z) \qquad \text{(Equation 5.4)}$$

The rational function of z multiplying the z-transform of the input $X(z)$ is the mathematical representation of the DSP system, called **T(z)** or **the transfer function**, since if you multiply it by the input z-transform, you get the output z-transform. In subsequent chapters this representation of the DSP system will be used to analyze, design, and represent digital filters.

Example 5.4. Finding the transfer function of a lowpass digital filter

Problem: From Chapter 1 we have a crude lowpass digital filter given by the following equation. We would like to get the transfer function $T(z)$ of this lowpass filter.

$$y = 0.5x + 0.5x_{-1}$$

Solution: The modified DSP equation, as described in the preceding section, is given in the following equation and then the z-transform of this equation is taken.

$$y(n) = 0.5x(n) + 0.5x(n-1)$$

The z-transform of the preceding equation is given in the following equation.

$$Y(z) = 0.5X(z) + 0.5z^{-1}X(z)$$

This equation can be easily solved for $Y(z)$ in terms of $X(z)$ to get the following equation.

$$Y(z) = (0.5 + 0.5z^{-1})X(z)$$

Thus it can be seen that the transfer function $T(z)$ of the lowpass digital filter is given by the preceding equation after solving for $Y(z)$ over $X(z)$.

$$T(z) = 0.5 + 0.5z^{-1}$$

Usually the transfer function is put in a more standard form by using algebra to eliminate negative exponents of z, as is shown in Example 5.5.

Example 5.5. Finding the transfer function of a more complex digital filter

Problem: The difference equation of a digital filter is given in the following equation. We want to find the mathematical description of this filter $T(z)$.

$$y(n) = 1.85y(n-1) - 0.868y(n-2) + 0.00477x(n-1) + 0.00455x(n-2)$$

Solution: Taking the z-transform of all the signals in the equation, we get the following equation.

$$Y(z) = Y(z)[1.85z^{-1} - 0.868z^{-2}] + X(z)[0.00477z^{-1} + 0.00455z^{-2}]$$

The z-Transform of the DSP Equation

Solving this for $T(z)$ gives the following equation.

$$T(z) = \frac{Y(z)}{X(z)} = \frac{0.00477z^{-1} + 0.00455z^{-2}}{1 - 1.85z^{-1} + 0.868z^{-2}}$$

If the numerator and denominator of the preceding equation are both multiplied by z^2, we get the following equation for $T(z)$, which is actually the same as the preceding equation.

$$T(z) = \frac{0.00477z + 0.00455}{z^2 - 1.85z + 0.868}$$

As we have shown, the z-transform allows us to get a mathematical description called $T(z)$ or z-transfer function of a digital filter or DSP system from the difference equation related to the coding. This transfer function will be solved by the methods in Chapter 6 to give the frequency response or gain of the digital filter. The methods used here could also be used in reverse order to get the difference equation to code. This is illustrated in Example 5.6. This procedure will be used in subsequent chapters when we determine the digital filter transfer function to meet the graphical specifications. The filter will obviously be given as $T(z)$, which is the digital filter expressed mathematically.

Example 5.6. Finding the difference equation from T(z)

Problem: Given the following math description of a digital filter, determine the corresponding difference equation that could easily be coded.

$$T(z) = \frac{0.25z^2 - 0.5z + 0.25}{z^2 - 0.95z + 0.75}$$

Solution: First let's divide numerator and denominator by z^2, as shown in the following equation.

$$T(z) = \frac{Y(z)}{X(z)} = \frac{0.25 - 0.5z^{-1} + 0.25z^{-2}}{1 - 0.95z^{-1} + 0.75z^{-2}}$$

Then cross-multiply the preceding equation to get the following equation.

$$Y(z)[1 - 0.95z^{-1} + 0.75z^{-2}] = X(z)[0.25 - 0.5z^{-1} + 0.75z^{-2}]$$

Now multiplying through each term in the brackets gives the following equation.

$$Y(z) = 0.95\,Y(z)z^{-1} - 0.75\,Y(z)z^{-2} + 0.25X(z) - 0.5X(z)z^{-1} + 0.75X(z)z^{-2}$$

Finally the inverse z-transform of the signals can be taken by applying the shifting property in reverse order.

$$y(n) = 0.95y(n-1) - 0.75y(n-2) + 0.25x(n) - 0.5x(n-1) + 0.25x(n-2)$$

Summary

Several important things done in this chapter will be necessary for digital filter analysis and design in following chapters. The first thing was to rewrite the general DSP equation in a more mathematical and standard form, as given by Equation 5.2. The second thing done was to use the z-transform property in Chapter 4 on the DSP equation given in Equation 5.2. Finally this z-transformed equation was solved to give the transform of the output over the input, which is the transfer function $T(z)$. Since multiplying the transform of the input by $T(z)$ gives the transform of the output, it must be a mathematical description of the DSP system. Using this mathematical description, we will be able to analyze, design, and represent digital filters in subsequent chapters.

Self-Test

1. Given the simple difference equation for digital integration given in Chapter 1 and repeated here, convert it into the more mathematical and standard modified difference equation using the steps given in Section 5.1.

$$y = y_{-1} + Tx$$

2. Using the steps given in Section 5.1, convert the following difference equation, given in the form of Chapter 1, into the more mathematical and standard form.

$$y = a_{-1}y_{-1} + bx + b_{-1}x_{-1}$$

3. Given the following difference equation of a digital filter, find the transfer function $T(z)$.

The z-Transform of the DSP Equation

$$y(n) = -2y(n-1) + 3x(n) + x(n-1)$$

4. Given the following difference equation of a digital filter, find the transfer function $T(z)$.

$$2y(n) + y(n-1) = -3y(n-2) + 4x(n) - 2x(n-2)$$

5. Given the following transfer function T(z) of a DSP system, write the difference equation.

$$T(z) = \frac{3 + 2z^{-1} + z^{-2}}{1 - 4z^{-1} + 5z^{-2}}$$

6. Given the following transfer function for $T(z)$ of a digital filter, write the difference equation.

$$T(z) = \frac{a_0 + a_1 z^{-1}}{1 - b_1 z^{-1}}$$

7. Determine the difference equation from the following transfer function $T(z)$.

$$T(z) = \frac{3z + 5}{2z^2 - 5z + 4}$$

8. Rewrite the transfer function in Problem 4 so that it has no negative exponents.

9. Rewrite the transfer function in Problem 5 so that it has no negative exponents.

10. Write the difference equation corresponding to the transfer function $T(z)$ determined in Problem 8.

11. Write the difference equation corresponding to the transfer function $T(z)$ determined in Problem 9.

12. Determine the difference equation for the following highpass digital filter, with the latest output sample being $y(n)$.

$$T(z) = \frac{0.8(z-1)}{z-0.6}$$

13. Determine the difference equation for the following lowpass digital filter, with the latest output sample being $y(n)$.

$$T(z) = \frac{0.2(z+1)}{z-0.6}$$

14. Determine the difference equation for the following second-order Butterworth digital filter, with the latest output sample being $y(n+2)$.

$$T(z) = \frac{0.04414(z+1)(z+1)}{z^2 - 1.324z + 0.5006}$$

Problems

1. Given the following simple difference equation for digital integration, convert it into the more mathematical and standard difference equation using the steps given in Section 5.1.

$$y = y_{-1} + \frac{T}{2}(x + x_{-1})$$

2. Using the steps given in Section 5.1, convert the following difference equation into the more mathematical and standard form.

$$y = a_{-1}y_{-1} + a_{-2}y_{-2} + b_{-1}x_{-1}$$

3. Given the following difference equation of a digital filter, find the transfer function $T(z)$.

$$y(n) = -4y(n-1) - 2y(n-2) + 7x(n)$$

4. Given the following difference equation of a digital filter, find the transfer function $T(z)$.

$$-3y(n-1) + y(n) = 2x(n-1) - y(n-2)$$

5. Given the following transfer function $T(z)$ of a DSP system, write the difference equation.

$$T(z) = \frac{4 - 2z^{-1} + 3z^{-2}}{1 - 6z^{-1} - 3z^{-2}}$$

6. Given the following transfer function for $T(z)$ of a digital filter, write the difference equation.

$$T(z) = \frac{2a_0 + 0.5a_1 z^{-1}}{1 - b_1 z^{-1} - b_2 z^{-2}}$$

7. Determine the difference equation from the following transfer function.

$$T(z) = \frac{5z + 2}{z^2 - 6z - 1}$$

8. Rewrite the transfer function in Problem 4 so that it has no negative exponents.

9. Rewrite the transfer function in Problem 5 so that it has no negative exponents.

10. Write the difference equation corresponding to the transfer function $T(z)$ determined in Problem 8.

11. Write the difference equation corresponding to the transfer function $T(z)$ determined in Problem 9.

12. Determine the difference equation for the following highpass digital filter, with the latest output sample being $y(n)$.

$$T(z) = \frac{0.677(z - 1)}{z - 0.987}$$

13. Determine the difference equation for the following lowpass digital filter, with the latest output sample being $y(n)$.

$$T(z) = \frac{0.4(z + 1)}{z - 0.878}$$

14. Determine the difference equation for the following second-order digital filter, with the latest output sample being $y(n + 2)$.

Digital Signal Processing

$$T(z) = \frac{0.124(z+1)(z+1)}{z^2 - 15.28z + 0.878}$$

Answers to Self-Test

1. $y(n) = y(n-1) + Tx(n)$

2. $y(n) = a_1 y(n-1) + b_0 x(n) + b_1 x(n-1)$

3. $T(z) = \dfrac{3 + z^{-1}}{1 + 2z^{-1}}$

4. $T(z) = \dfrac{4 - 2z^{-2}}{2 + z^{-1} + 3z^{-2}}$

5. $y(n) = 4y(n-1) - 5y(n-2) + 3x(n) + 2x(n-1) + x(n-2)$

6. $y(n) = b_1 y(n-1) + a_0 x(n) + a_1 x(n-1)$

7. $y(n) = 2.5y(n-1) - 2y(n-2) + 1.5x(n-1) + 2.5x(n-2)$

8. $T(z) = \dfrac{4z^2 - 2}{2z^2 + z + 3}$

9. $T(z) + \dfrac{3z^2 + 2z + 1}{z^2 - 4z + 5}$

10. $y(n+2) = -0.5y(n+1) - 1.5y(n) + 2x(n+2) - x(n)$

11. $y(n+2) = 4y(n+1) - 5y(n) + 3x(n+2) + 2x(n+1) + x(n)$

12. $y(n) = 0.6y(n-1) + 0.8x(n) - 0.8x(n-1)$

13. $y(n) = 0.6y(n-1) + 0.2x(n) + 0.2x(n-1)$

14. $y(n+2) = 1.324y(n+1) - 0.5006y(n) + 0.04414[x(n+2) + 2x(n+1) + x(n)]$

chapter 6

Frequency Response of Digital Filters and DSP Systems

Introduction

In this chapter we learn to determine the frequency response of any DSP system from the system transfer function $T(z)$. By doing this we will be able to determine if a digital filter meets the requirements of the graphical filter specifications of Chapter 3 and how well it meets the specifications. This analysis needs to be done on any DSP system to check the design, and it also gives a deeper insight into digital filters and how they work. In order to obtain the frequency response, we will show graphically and mathematically how the z-transform variable is related to the Laplace transform variable of continuous signals, and how a simple mathematical substitution for this variable gives the frequency response. In order to make the computation easier on most calculators and some mathematical programs, we will introduce the Euler equation. If the student has not used Laplace transforms in analog signal processing or analog control classes, the student can still use the resulting trigonometric substitution into $T(z)$ to obtain the frequency result.

6.1 The Euler Equation from Trigonometry

Many calculators and mathematical programs cannot compute the expression $e^{j\Omega}$, where j is the square root of -1 and Ω is an angle in radians. The computation will occur frequently in finding the frequency response of digital filters, and the reason will soon be shown. In order to ease the computation burden, the **Euler** equation given in Equation 6.1 can be used.

$$Ae^{j\Omega} = A\cos(\Omega) + jA\sin(\Omega) \qquad \text{(Equation 6.1)}$$

The expression on the right is just a complex number in rectangular form with magnitude A and phase Ω. Thus the expression on the left must be a complex number in polar form with magnitude A and angle Ω. This relationship is shown in Figure 6.1.

The student may be used to seeing complex numbers in polar form given as $A\angle\Omega$, but this form is just shorthand mathematical symbols saying magnitude A at an angle of Ω, not real mathematics. You could double the A value to express a complex number twice as big, but you cannot use this shorthand math to take the integral or derivative of this complex number if it were variable. The correct mathematical expression of a complex number in polar form with magnitude A at an angle of Ω is given on the left side of Equation 6.1. As all calculators and programming languages and mathematical applications programs can evaluate the expression on the right, it will be used in the rest of this text in evaluating the frequency response of digital systems. Example 6.1 gives an example of using the Euler equation.

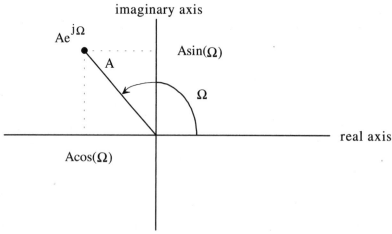

Figure 6.1. Exponential and rectangular forms of a complex number

Frequency Response of Digital Filters and DSP Systems

Example 6.1. Using the Euler equation

Problem: Find the numeric value for $7e^{-1.2j}$.

Solution: Using Equation 6.1, we have the following result.

$$7e^{-1.2j} = 7\cos(-1.2) + j7\sin(-1.2)$$

$$= 7\cos(1.2) - j\sin(1.2)$$

$$= 2.54 - j6.52$$

$$= 7.00 \angle -1.2 \text{ rad.}$$

6.2 Frequency Scaling

We will also see when we try to compute the frequency response of digital filters that the analog input signal frequencies w used in the graphical specifications will always be multiplied by the sampling period T. It is usual practice in industry to define Ω as wT. This is just a new symbol that represents the input signal frequency multiplied by the sampling period T, called the **scaled frequency**. When this new symbol is used, the frequency response of digital filters inherently goes from 0 to π rad/s before repeating, as shown in Equation 6.2. Remember we showed that any DSP system has a frequency response that repeats at half the sampling frequency $w_s/2$.

$$\frac{\Omega_s}{2} = \frac{w_s T}{2}$$

$$= \frac{1}{2}(\frac{2\pi}{T})$$

$$= \pi \qquad \text{(Equation 6.2)}$$

Almost all texts and technical articles will show DSP frequency responses plotted going from 0 to π, since they are in terms of $wT = \Omega$. In order to find out what the digital filter or DSP system does to the input frequencies from the plots, just divide the plot Ω by T to obtain the specific input frequency of the DSP system. Example 6.2 illustrates how to convert between the actual input or output frequency in radians and the parameter Ω usually used in texts and industry.

Example 6.2. Converting between the scaled frequency Ω and the actual frequency

Problem: A graphical specification of a digital filter gives its gain at $\Omega = 0.5$ as −6 dB with $T = 0.02$. Find the value of the input or output frequency at which the filter has this gain.

Solution: Since $wT = \Omega$, we have

$$w = \Omega/T$$

$$= 0.5/.02 = 10 \text{ rad/s}$$

6.3 Computing the DSP Frequency Response

In order to compute the frequency response of any DSP system, we must first determine the relationship of the z-transform variable z and the Laplace variable s. Once this is done, computing the frequency response is easy, since from analog filtering courses we know that replacing s by jw in an analog filter transfer function gives the frequency response. The relationship is easy to establish by remembering that if an analog continuous time signal is delayed by $t = T$, then its Laplace transform is multiplied by e^{-sT}. This Laplace transform property is shown by the following equation, where L signifies the Laplace transform of the term inside the brackets. This is called the shifting property of Laplace transforms.

$$L[x(t - T)] = e^{-sT} L[x(t)]$$

In Chapter 4 (Section 4.4) we also showed that when a discrete-time signal is delayed by T, then its z-transform is multiplied by z^{-1}. The equation for the shifting property of the z-transform is given in the following equation.

$$Z[x(nT - T)] = Z[x(n - 1)] = z^{-1} Z[x(n)]$$

As a discrete-time signal can be thought of as a continuous time signal

Frequency Response of Digital Filters and DSP Systems

with a signal value of zero between samples, we have the relationship given by Equation 6.3.

$$z^{-1} = e^{-sT} \rightarrow z = e^{sT} \qquad \text{(Equation 6.3)}$$

This is a very important relationship to remember.

With the relationship given in Equation 6.3, we could actually write the Laplace transfer function of any DSP system by replacing z with e^{sT} in $T(z)$. But more important, we can get the frequency response by letting $s = jw$, so that $z = e^{jwT}$ in $T(z)$ everywhere z appears. Since z appears everywhere in the z-transfer function and $z = e^{jwT}$, you can see why just for notational convenience $wT = \Omega$ is used as shown in Section 6.2, so that we have

$$z = e^{j\Omega} = \cos(\Omega) + j\sin(\Omega)$$

after using the Euler equation given in Section 6.1.

We now have the method of finding the frequency response magnitude (gain) as shown in Equation 6.4.

$$\text{gain} = |T(z)|_{z = \cos(\Omega) + j\sin(\Omega)}$$

$$\text{gain}_{dB} = 20\log(|T(z)|_{z = \cos(\Omega) + j\sin(\Omega)}) \qquad \text{(Equation 6.4)}$$

In Equation 6.4, $\Omega = wT$, where T is the period between samples of the input or output. Since in digital filtering we are most interested in the magnitude of the frequency response, as stated in Chapter 3, we need to take the magnitude, as the absolute value symbols indicate.

Equation 6.4 shows that the frequency response of all DSP systems repeat every w_s, since Ω_s is given by

$$\Omega_s = (2\pi/T)T = 2\pi$$

and the trigonometric functions in Equation 6.4 repeat every 2π. Thus the value of z is unchanged in Equation 6.4 if any frequency is increased by integer multiples of the sampling frequency, as was stated in Chapter 2.

Equation 6.4 looks like a pretty easy equation to solve for the gain as a function of Ω, but on a calculator even a few steps soon become tiresome finding the gain for just one value of Ω. It is usually programmed on a calculator, or a math package such Mathcad is used. In Example 6.3, Equation 6.4 is used to find the gain of a simple lowpass filter showing the steps needed if a calculator is used. In Example 6.4 we show how to calculate the gain and plot the gain for several frequencies using Mathcad. Finally, in Example 6.5, the gain for a filter that uses both input and output samples is calculated at one frequency.

Example 6.3. Finding the gain for one value of input frequency using a calculator

Problem: Find the magnitude of the frequency response (gain) of the following lowpass filter by using Equation 6.4, for $w = 50$ rad/s with the sample period $T = 0.01$ s.

$$y(n) = 0.5x(n) + 0.5x(n-1)$$

Solution: In Chapter 5, the z-transfer function was shown to be

$$T(z) = \frac{Y(z)}{X(z)}$$

which can be obtained from the filter difference equation by taking the z-transform both sides and solving for the ratio, as is done in the following equations.

$$Y(z) = 0.5X(z) + 0.5z^{-1}X(z)$$

$$T(z) = 0.5 + 0.5\,z^{-1}$$

Using the preceding equation and the values for T and w in Equation 6.4 gives the gain as

$$\text{gain} = |T(e^{j\Omega})| = |0.5 + 0.5[\cos(\Omega) + j\sin(\Omega)]^{-1}|$$

Using the Euler equation, it is easy to see that

$$[\cos(\Omega) + j\sin(\Omega)]^{-1} = \cos(\Omega) - j\sin(\Omega)$$

since this just changes $e^{j\Omega}$ to $e^{-j\Omega}$ and the sine of a negative angle is just the negative of the sine of the corresponding positive angle. Thus we have the final answer for the gain after using the values for w and T as shown in the following equations.

Frequency Response of Digital Filters and DSP Systems

$$\Omega = wT = 0.5$$

$$\text{gain} = |0.5 + 0.5\cos(0.5) - j0.5\sin(0.5)|$$

$$= |(0.5 + 0.439) - j(0.24)|$$

$$= \sqrt{(0.5 + 0.439)^2 + (0.24)^2}$$

$$= 0.969$$

Let's look at what has been done in Example 6.3. We have computed the gain of a digital filter (it could have been any DSP system) using the difference equation of the filter that would be coded on a computer or DSP chip. The gain was computed for the frequency of 50 rad/s at a sample period of 0.01 s. What this means is that for any input sinusoidal signal or sinusoidal component of a signal going into the ADC with a frequency of 50 rad/s, the output amplitude of the sinusoid at the same frequency would be 0.969 times as great coming out of the DAC. By solving for many input frequency points, we can determine the effect of the filter on any input frequency. Note that if the sample time T is changed, Equation 6.4 and Example 6.3 show that new and different values of gain are obtained. Thus the frequency response of a digital filter varies with the sample period used.

Example 6.4. Using Mathcad to calculate and plot the gain in Example 6.3

Problem: The filter difference equation and sampling period are the same as in Example 6.3, but now we would like to draw the gain curve by computing enough gain values at different frequencies so that a plot of the gain can be drawn.

Solution:

$n := 1..50$ The number of points to calculate

$w_n := 10 \cdot n$ The frequency points in rad/s

$T := 0.01$ The sample period in s

$z_n := \cos(w_n \cdot T) + \sin(w_n \cdot T) \cdot j$ The value of z for every frequency calculated

Digital Signal Processing

$$\text{gain}_n := \left| 0.5 + 0.5 \cdot (z_n)^{-1} \right|$$

The magnitude of the freq. resp. at each freq.

$$\text{gaindB}_n := 20 \cdot \log(\text{gain}_n)$$

The gain in dB at each freq. point

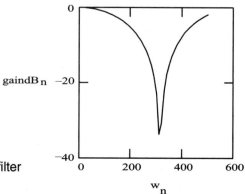

Figure 6.2. Gain plot of lowpass filter

Example 6.5. Calculating the gain at one frequency for a filter with a and b coefficients

Problem: Given the following digital filter equation, find its gain for an input frequency or output frequency of $w = 100$ rad/s, for $T = 0.0126$ s between input samples.

$$y(n) = 0.7y(n-1) + x(n)$$

Solution: Taking the z-transform of both sides of the preceding equation gives

$$Y(z) = 0.7z^{-1}Y(z) + X(z)$$

Solving this for $T(z)$, the transfer function of the digital filter gives

$$T(z) = \frac{1}{1 - 0.7z^{-1}}$$

Using this equation in Equation 6.4 gives the following equation for the gain after using the values for w and T.

$$\Omega = wT = 1.26$$

Frequency Response of Digital Filters and DSP Systems

$$\text{gain} = \left| \frac{1}{1 - 0.7[\cos(1.26) - j\sin(1.26)]} \right|$$

$$= \left| \frac{1}{1 - 0.7(0.306 - j0.952)} \right|$$

$$= \left| \frac{1}{1 - 0.212 + j0.666} \right|$$

$$= \left| \frac{1}{0.786 + j0.666} \right|$$

$$= 0.939$$

Summary

In Chapter 6 we learned how to find the gain, which is the magnitude of the frequency response, of any DSP system, given its mathematical description $T(z)$. The method is given in Equation 6.4 and illustrated in Examples 6.3, 6.4, and 6.5. In order to make the notation and computations easier, Equation 6.1 shows the conversion from the polar or exponential form of a complex number, which occurs naturally when finding the frequency responses of DSP systems, to the more familiar rectangular form. Also the definition of Ω as wT was introduced, since it is used a lot in industry and other texts, and simplifies the notation. Using these last two formulas, it is easy to compute the DSP system gain in terms of the input or output frequency w, given the sampling period T, by using Equation 6.4.

Self-Test

1. For the following complex numbers in exponential form, determine their values in rectangular coordinates.

    (a) $3e^{-4j}$

    (b) $-8e^{0.5j}$

2. For the following input frequencies, determine the corresponding scaled frequency Ω for the given sample period T.

(a) $w = 100$ rad/s, $T = 0.02$ s

(b) $w = 2000$ rad/s, $T = 0.0003$ s

3. For the following difference equation of a digital filter, determine the gain at $w = 6$ rad/s with $T = 0.05$ s.

$$y(n) = 0.905y(n-1) + 0.119x(n)$$

4. For the following difference equation of a digital filter, determine the gain at $w = 4$ rad/s with $T = 0.1$ s.

$$y(n+1) + 0.6y(n) + 10x(n+1) - 10x(n)$$

5. A first-order highpass digital filter transfer function is given here for $T = 0.01$ s. Find the frequency response magnitude (gain in dB) at 50 and 100 rad/s.

$$T(z) = \frac{0.8(z-1)}{z-0.6}$$

6. Determine the gain in dB at 5 and 20 rad/s for the digital filter with the following transfer function with the sampling period $T = 0.05$ s. Also state if it is a highpass or lowpass digital filter.

$$T(z) = \frac{0.2(z+1)}{z-0.6}$$

7. Use Mathcad to plot the gain of the digital filter with the following transfer function $T(z)$ for a sampling period of $T = 0.005$. The filter is a second-order Butterworth with the corner frequency at 100 rad/s. Plot the gain versus frequency in rad/s.

$$T(z) = \frac{0.04414(z+1)(z+1)}{z^2 - 1.324z + 0.5006}$$

8. Repeat Problem 7 except plot the gain in dB and the frequency in terms of $\Omega = wT$.

9. Determine the gain in dB at 2 rad/s and 37 rad/s for the digital filter

with the following transfer function with the sampling period $T = 0.02$ s. Also state if it is a highpass or lowpass filter.

$$T(z) = \frac{0.8333(z-1)}{z - 0.667}$$

10. Determine the gain in dB at 2 rad/s and 37 rad/s for the digital filter with the following transfer function with the sampling period $T = 0.01$ s. Also state if it is a highpass or lowpass filter.

$$T(z) = \frac{0.9091(z-1)}{z - 0.8182}$$

11. For the digital filter with the transfer function in Problem 5, determine the gain in dB at 100 rad/s if the sampling period $T = 0.005$ s.

12. For the digital filter with the transfer function in Problem 10, determine the gain in dB at 2 rad/s if the sampling period $T = 0.185$ s.

13. For the following difference equation of a digital filter, use Mathcad to compute the gains at $w = 3, 6, 12,$ and 24 rad/s for $T = 0.025$ s.

$$y(n+1) = 0.839y(n) + 0.0805x(n+1) + 0.0805x(n)$$

14. For the following equation of a digital filter, use Mathcad to compute the gains at $w = 2, 4, 8, 16$ rad/s for $T = 0.08$ s.

$$y(n+1) = 0.667y(n) + 4.167x(n+1) - 4.167x(n)$$

15. For the following first-order lowpass filter $T(z)$, use Mathcad to compute the gains at $w = 10, 20, 40, 80, 160$ rad/s for $T = 0.02$ s.

$$T(z) = \frac{0.4(z+1)}{z - 0.6}$$

16. For the following first-order digital filter $T(z)$, use Mathcad to plot the gain in dB and state if the filter is a highpass or lowpass filter for $T = 0.01$ s.

$$T(z) = \frac{0.677(z-1)}{z - 0.333}$$

Digital Signal Processing

Problems

1. For the following complex numbers in exponential form, determine their values in rectangular coordinates.

   (a) $-7e^{-0.5j}$

   (b) $0.25e^{4j}$

2. For the following input frequencies, determine the corresponding scaled frequency Ω for the given sample period T.

   (a) $w = 33.5$ rad/s, $T = 0.005$ s

   (b) $w = 400$ rad/s, $T = 2.5$ s

3. For the following difference equation of a digital filter, determine the gain at $w = 12$ rad/s with $T = 0.007$ s.

$$y(n+1) = 0.915y(n) + 0.5x(n+1)$$

4. For the following difference equation of a digital filter, determine the gain at $w = 45$ rad/s with $T = 0.02$ s.

$$y(n+2) = 0.85y(n+1) + 5x(n+1) + x(n)$$

5. A first-order highpass digital filter transfer function is given here for $T = 0.005$ s. Find the frequency response magnitude (gain in dB) at 75 and 125 rad/s.

$$T(z) = \frac{0.75(z-1)}{z-0.7}$$

6. Determine the gain in dB at 10 and 30 rad/s for the digital filter with the following transfer function with the sampling period $T = 0.001$. Also state if it is a highpass or lowpass digital filter.

$$T(z) = \frac{0.5(z-1)}{z-0.6}$$

7. Use Mathcad to plot the gain of the digital filter with the following

transfer function $T(z)$ for a sampling period of $T = 0.002$ s. The filter is a second-order Chebyshev with the corner frequency at 100 rad/s. Plot the gain versus frequency in rad/s.

$$T(z) = \frac{0.00661(z+1)(z+1)}{z^2 - 1.853z + 0.8796}$$

8. Repeat Problem 7 except plot the gain in dB and the frequency in terms of $\Omega = wT$.

9. Determine the gain in dB at 15 rad/s and 30 rad/s for the digital filter with the following transfer function with the sampling period $T = 0.01$ s. Also state if it is a highpass or lowpass filter.

$$T(z) = \frac{0.667(z-1)}{z - 0.333}$$

10. Determine the gain in dB at 15 rad/s and 30 rad/s for the digital filter with the following transfer function with the sampling period $T = 0.01$ s. Also state if it is a highpass or lowpass filter.

$$T(z) = \frac{0.0476(z+1)}{z - 0.905}$$

11. For the digital filter with the transfer function in Problem 5, determine the gain in dB at 150 rad/s if the sampling period $T = 0.0025$ s.

12. For the digital filter with the transfer function in Problem 10, determine the gain in dB at 5 rad/s if the sampling period $T = 0.005$ s.

Answers to Self-Test

1a. $-1.961 + 2.270j$

1b. $-7.021 - 3.835j$

2a. 2

2b. 0.6

3. 0.397

4. 7.8

5. −2.9 dB and −0.83 dB respectively

6. −1 dB, −8 dB, lowpass

7. w = 100 rad/s; gain dB = −3.2; w = 200 rad/s; gain dB = −13.8

8. w = 100 rad/s, Ω = 0.5; w = 200 rad/s, Ω = 1.0

9. −20 dB, −1 dB, highpass

10. −20dB, −1dB, highpass

11. −2.9 dB

12. −1 dB

13. 0.92, 0.75, 0.50, 0.27

14. 1.86, 3.14, 4.27, 4.83

15. 1.86, 1.55, 1.02, 0.47, 0.01

16. highpass, corner frequency at w = 100 rad/s

chapter 7

IIR Filter Design

Introduction

In this chapter we show how to design one of the two types of digital filters: the **Infinite Impulse Response (IIR)** filter. This digital filter has a gain curve that approximates the filter characteristics of a corresponding analog filter. There are four basic analog filter approximations themselves, which are reviewed in Section 7.1. One of the best ways to obtain the digital filter approximations to analog filters is to develop a mathematical formula that shows how to convert from $T(s)$, the mathematical description of the analog filter using Laplace transforms, to $T(z)$, the description of the digital filter. Then, using the methods in Chapter 5 in reverse order, obtain the coding for a corresponding IIR filter.

IIR filters are used in many areas of technology, some of which are listed here. Application 1 at the end of this chapter shows how to design, analyze, and determine the difference equation to code as well as a flowchart to illustrate the coding procedure. Because IIR filters approximate the gain and phase response of analog filters, they are used primarily where analog filters are used. However, implementation on a processor allows much more flexibility, eliminates degradation, and produces a specific accuracy based on the number of bits used, as well as perfect filter reproducibility. Some of the applications areas are for sound and music enhancement, telecommunications, video image processing, biomedical instrumentation, and radar and sonar processing. (Digital control systems use IIR filters as compensators, but the IIR filter design methods in this and any digital filter text are not appropriate when applied to control systems with feedback.)

Unfortunately, there are many mathematical formulas that approximate an analog filter with an IIR filter, since a computer or DSP chip can only do addition, subtraction, multiplication, and division; everything else is an approximation. There are many ways to approximate a function like a filter, so there are many ways to get IIR filters. Some have the advantage of simplicity, and others are more accurate for the same number of coefficients. We will look at three approximation methods in Sections 7.2, 7.3, and 7.4. These formulas will start with the Laplace transform $T(s)$ of the analog filter that is to be replaced by a digital IIR filter, then use algebra and the relationship between the z-transform and the Laplace transform to obtain the IIR filter that has the same approximate outputs at the sample times. Remember, there is one fundamental difference that cannot be approximated away: The IIR filter will have its gain start to repeat the lower frequency gain above half the sampling frequency! Most gain plots of digital filters go from zero or some low frequency up to $wT = \Omega = \pi$.

7.1 Review of Four Basic Analog Filter Approximations

Many types of analog filters can be built. Any one could be a lowpass, highpass, bandpass, or a stopband filter like those discussed in Chapter 3. However, because of the nature of electrical circuits used to build analog filters, any of these filter types can be divided into four basic analog approximations that meet the graphical specification. These approximations are based on where the gain curve has **ripples** or deviations from a smoothly varying curve. In the first approximation, called the **Butterworth**, there are no ripples in any passband or stopband. Thus the digital IIR filter has no ripples in it either. The general gain curve is given in Figure 7.1 for a lowpass filter specification. Similar graphical specifications could be drawn for highpass, bandpass, or bandstop filters. Notice in Figure 7.1 that the important characteristic is that the gain curve smoothly varies in the passband and the stopband up to half the sampling frequency.

The second analog filter approximation to an ideal analog graphical filter specification is the **Chebyshev**, which has ripples in the passbands, but has a smoothly decreasing gain curve in the stopbands. In Figure 7.2 a lowpass

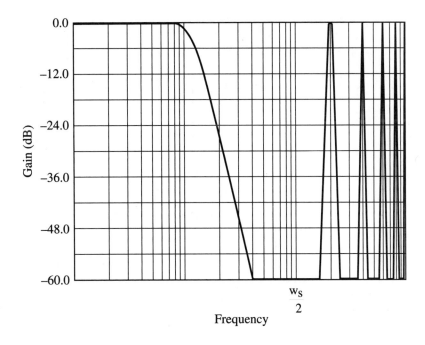

Figure 7.1 Butterworth lowpass IIR filter

IIR filter is used to illustrate a Chebyshev approximation. Similar gain curves could be drawn for a highpass, bandpass, or bandstop filter. Notice in Figure 7.2 that the gain curve has ripple in the passband, because the gain increases before it decreases. For higher-order filters the ripple is more obvious, with several cycles of increasing and decreasing gain in the passband. This is an unwanted deviation from the ideal analog filter. However, as the Chebyshev filter will have a narrower transition band between the stop and passbands, it trades off ripple in the passband for a gain curve that more closely approximates the ideal graphical specification by having a narrower transition band than the Butterworth filter.

The third analog filter approximation to the ideal analog graphical filter specification is the **Inverse Chebyshev**, which has no ripple in the passbands, but has ripple in the stopbands. In Figure 7.3, an IIR bandpass filter is used to illustrate a digital approximation to an analog Inverse Chebyshev bandpass filter. Again, the ripples are an unwanted deviation from the ideal graphical specifications, but like the Chebyshev approximation it has a narrower transition band than the Butterworth. Notice in Figure 7.3 that the passband has no ripples but there are ripples in the

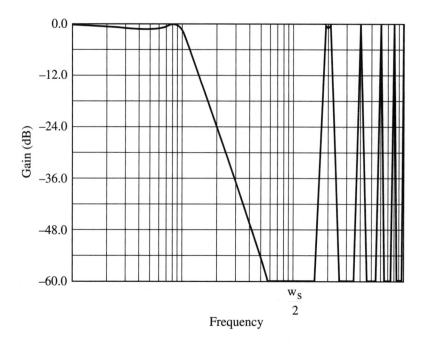

Figure 7.2. Chebyshev lowpass IIR filter

stopbands. Many times this ripple in the stopbands is insignificant if the peaks of the ripples are under the stopband gain specification. This is in contrast to the Chebyshev filter, where the ripple in the passband is a deviation from the desired gain curve.

The fourth type of analog filter approximation to the ideal analog graphical specifications is the **Cauer**, which has ripple in the passbands and the stopbands. The reason this ripple is accepted is that the Cauer filter has narrower transition bands than any of the other three approximations. Figure 7.4 illustrates the digital IIR graphical specification for a bandstop filter. Notice the ripple in the stopband and the passbands. If a highly selective filter is desired and some ripple in the gain curve is acceptable, then the Cauer filter is the best choice.

7.2 The Impulse Invariant IIR Filter

One way to determine a digital IIR filter to approximate an analog filter of any of the four basic approximations given in Section 7.1, is the

IIR Filter Design

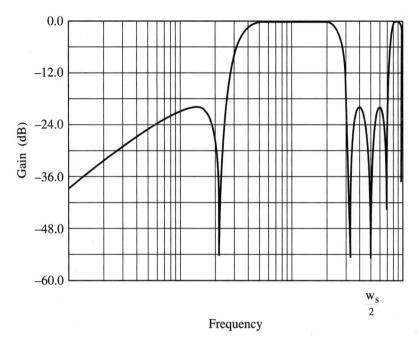

Figure 7.3. Inverse Chebyshev bandpass filter

Impulse Invariant method. It is obtained by solving for the $T(z)$ that has the same output values as the analog filter at the sample times when both have an impulse input. The formula for this method is derived below. It should be noted that the digital filter obtained by this method is an approximation for all inputs, and it is exact for an impulse input. This is the easiest IIR filter design method to use, but as it is most accurate for low frequencies, it is usually only used for lowpass filters.

For either the Laplace transform or the z-transform, the transform of the output is just the input transform multiplied by the corresponding transfer function, $T(s)$ or $T(z)$. This was shown in Chapter 4 for the z-transform and is given here along with the corresponding equation for the Laplace transform. Remember $Y(s)$ or $Y(z)$ is the transformed output for $X(s)$ or $X(z)$, the transformed input.

$$Y(s) = T(s)X(s)$$

$$Y(z) = T(z)X(z)$$

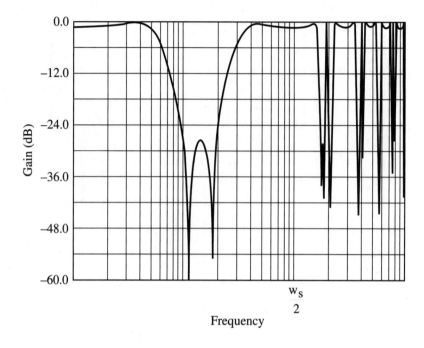

Figure 7.4. Cauer bandstop IIR filter

However, as the Laplace transform and the z-transform of a unit impulse are 1, the transforms of the outputs are

$$Y(s) = T(s)$$

$$Y(z) = T(z)$$

Now the analog filter output in the time domain is just the inverse Laplace transform, and if we replace t by nT, we have the output due to an impulse at the sample times, $y(nT)$, which is also written as $y(n)$ for notational convenience. This discrete-time signal can now be z-transformed to get $T(z)$. These steps are shown in the next equations, where L^{-1} refers to the inverse Laplace transform operation: that is, it gives the signal whose Laplace transform is in the brackets.

$$y(t) = L^{-1}[Y(s)] = L^{-1}[T(s)]$$

$$y(n) = y(nT) = y(t)|_{t=sT}$$

IIR Filter Design

$$T(z) = Y(z) = Z[y(n)]$$

$$T(z) = Z[y(n)] = Z[y(nT)]$$

$$T(z) = Z\{L^{-1}[T(s)]_{t=nT}\}$$

The final equation says that the mathematical description of the digital IIR filter is the z-transform of the analog signal after sampling whose Laplace transform is $T(s)$. This equation is usually shortened by writing it as Equation 7.1. Equation 7.1 is not mathematically correct, but it is really just shorthand notation saying to z-transform the sampled inverse Laplace transform of T(s).

$$T(z) = Z[T(s)]*T \qquad \text{(Equation 7.1)}$$

Notice that a multiplying factor, or T, appears in Equation 7.1. This is because even though the Laplace and z-transforms of the unit impulses are 1, the impulses themselves are not the same. For analog signals the impulse has infinite value at $t = 0$ but an area of 1, while a discrete time impulse has value 1 at $t = 0T = 0$. To account for this difference, the multiplying factor of T is used.

Equation 7.1 giving the formula for determining the digital IIR filter for an analog filter still looks forbidding, but with the use of tables like Table 4.1 (page 44), the process is easy. All you need to do is find the time signal in column 1 that corresponds to the analog signal with the Laplace transform $T(s)$ of the analog filter. Column 2 gives the sampled time signal y(n) whose z-transform is in column 3. Example 7.1 illustrates this process for a simple analog filter.

Example 7.1. Finding the impulse invariant IIR filter to approximate a first-order analog filter

Problem: A first-order lowpass filter with dc gain of 1 and a corner frequency $w_c = 10$ rad/s is given here. It can easily be seen to be a lowpass analog filter by letting $s = jw$. For $w \ll w_c$, the gain is approximately 1, and for $w \gg w_c$, the gain decreases linearly with w. We want to find a digital filter that approximates this analog filter, using the impulse invariant method for a sample time $T = 0.1$.

$$T(s) = \frac{10}{s+10}$$

Solution: Equation 7.1 says we must first get the time domain impulse response of $T(s)$, then sample it as if it went through an ADC with $T = 0.1$, and finally take the z-transform of the sampled signal. This z-transform will be the z-transfer function of the approximate digital filter. These steps, which really go from column 1 to column 3 in Table 4.1, are shown in the following equations, along with the scaling multiplication by T.

$$T(z) = Z[T(s)] * T$$

$$T(z) = Z\left[\frac{10}{s+10}\right] * T$$

$$T(z) = Z[10e^{-10nT}] * T$$

$$= Z[10e^{-10n(0.1)}](0.1) = Z[10(e^{-1})^n](0.1)$$

$$= \frac{10(0.1)z}{z - e^{-1}}$$

$$= \frac{z}{z - e^{-1}}$$

The gain curve of this digital filter is plotted in Figure 7.5 using the method given in Chapter 6. The repetition of the gain above half the

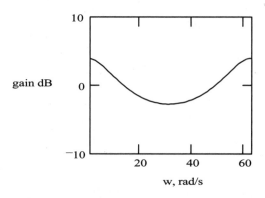

Figure 7.5 Gain plot of digital lowpass filter of Example 7.1

sampling frequency of 31.4 rad/s can be seen, as well as the lowpass filter characteristics with the gain dropping above 10 rad/s. In this example the poor approximation characteristics of the impulse invariant method can be seen for higher frequencies, since the sampling frequency was not well above the corner frequency.

In order to use the Impulse Invariant method for more complex analog filter transfer functions, it may be necessary to do a **Partial Fraction Expansion (PFE)** of $T(s)$ using algebra. After a PFE is done, all that needs to be done is to determine the inverse Laplace transform of each partial fraction in the expansion of $T(s)$ in column one of Table 4.1, and replace it by the corresponding z-transform in column 3. This procedure is shown in Example 7.2.

Example 7.2. Finding the IIR approximation to a more complex analog filter

Problem: Find the digital IIR filter for the second-order lowpass analog filter given by the following equation for $T(s)$ for a sampling period $T = 0.05$ s.

$$T(s) = \frac{15}{s(s+15)}$$

Solution: The first thing is to do a PFE of $T(s)$, so that it is the sum of terms that have time responses in column 1 of Table 4.1. This is shown in the following equations. Remember, to do a PFE without a mathematical package that does it, all you need to do is reverse putting $T(s)$ over a common denominator and then use algebra to solve for the unknown coefficients A_1 and A_2 in the sum of terms.

$$T(s) = \frac{15}{s(s+15)} = \frac{A_1}{s} + \frac{A_2}{s+15}$$

$$= \frac{A_1}{s} + \frac{A_2}{s+15}$$

Putting the terms back over a common denominator gives the following equation, which is easily solved for the numerator coefficients.

$$\frac{A_1(s+15) + A_2 s}{s(s+15)} = \frac{15}{s(s+15)}$$

$$A_1(s+15) + A_2 s = 15$$

$$s(A_1 + A_2) + 15A_1 = 15$$

From the preceding equation it is easy to see that

$$A_1 = 1$$

$$A_2 = -1$$

So we have

$$T(s) = \frac{1}{s} + \frac{-1}{s+15}$$

From Table 4.1 (page 44), we derive the following digital IIR filter equation.

$$T(z) = [\frac{z}{z-1} - \frac{z}{z-e^{-15T}}] * T$$

$$T(z) = 0.05 * \frac{z(1-e^{-0.75})}{(z-1)(z-e^{-0.75})}$$

$$T(z) = \frac{0.05z(1-0.472)}{(z-1)(z-0.472)}$$

$$T(z) = \frac{0.05z - 0.0264}{z^2 - 1.472z + 0.472}$$

We could use the procedure given in Chapter 6 to plot the gain curve of the preceding equation, as was done in Example 7.1. But this time let's use the methods given in Chapter 5 to derive the difference equation for the filter, which as was mentioned in Chapter 1 is easily converted to computer code. The following equations show the derivation of the filter difference equation.

$$T(z) = \frac{Y(z)}{X(z)} = \frac{0.05z - 0.0236}{z^2 - 1.472z + 0.472}$$

$$Y(z)(z^2 - 1.472z + 0.472) = X(z)(0.05z - 0.0236)$$

IIR Filter Design

$$Y(z)(1 - 1.472z^{-1} + 0.472z^{-2}) = X(z)(0.05z^{-1} - 0.0236z^{-2})$$

$$y(n) - 1.472y(n-1) + 0.472y(n-2) = 0.05x(n-1) - 0.0236x(n-2)$$

$$y(n) = 1.472y(n-1) - 0.472y(n-2) + 0.05x(n-1) - 0.0236x(n-2)$$

This last equation could easily be coded to approximate the analog filter, since it says the current output $y(n)$ is a weighted sum of two previous outputs and two previous inputs.

7.3 The Step Invariant IIR Filter

Another digital IIR filter approximation to an analog filter is found by using the **step invariant method**. This is usually a better approximation than the impulse invariant filter. The digital filter is exact at the sample times for a piecewise constant input, that is, one composed of discrete steps, as shown in Figure 7.6. The step invariant IIR filter is less exact as the input signal into the ADC deviates from a piecewise constant form. However, this is usually a better approximation to any input than a bunch of impulses, as is done by the impulse invariant approximation.

The step invariant method solves for the $T(z)$ that gives the same sample values as the analog filter $T(s)$ when both have step inputs. For the digital filter the input is $u(n)$, and for the analog filter the input is $u(t)$. The z-transform and Laplace transforms of these inputs are given and used in

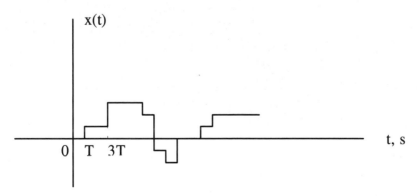

Figure 7.6. Example of piecewise constant analog input signal

the following equations to get the corresponding transformed output signals.

$$L[u(t)] = \frac{1}{s}$$

$$Z[u(n)] = \frac{z}{z-1}$$

$$Y(z) = T(z)U(z) = T(z)\frac{z}{z-1}$$

$$Y(s) = T(s)U(s) = \frac{T(s)}{s}$$

The output of the analog filter is $y(t)$, which is the inverse Laplace transform of $Y(s)$ in the preceding equation. If it were sampled every T seconds, it would be $y(n)$, which is the inverse z-transform of $Y(z)$ in the preceding equations. These signals can be used to solve for the z-transform of a digital filter with the same outputs at the sample times as the analog filter, since they are equated. The following equations show the solution for $T(z)$ given by Equation 7.2, which is the formula for an approximation to an analog filter using the step invariant method.

$$T(z) = \frac{z-1}{z}Y(z)$$

$$T(z) = \frac{z-1}{z}Z[y(n)]$$

$$T(z) = \frac{z-1}{z}Z[Y(s)]$$

$$T(z) = \frac{z-1}{z}Z[\frac{T(s)}{s}] \quad \text{(Equation 7.2)}$$

Remember that $Z[Y(s)]$ is just shorthand notation for the z-transform of the sampled inverse transformed signal $Y(s)$. Example 7.3 shows how Equation 7.2 is used to get a different digital IIR filter to approximate the analog filter in Example 7.1.

IIR Filter Design

Example 7.3. Finding the Step Invariant IIR filter to approximate a first-order lowpass analog filter

Problem: Find the transfer function $T(z)$ for a digital filter that approximates the analog lowpass filter in the following equation, using the step invariant method with the sample period $T = 0.1$ s.

$$T(z) = \frac{10}{s+10}$$

Solution: Using Equation 7.2, we have the following equations.

$$T(z) = \frac{z-1}{z} Z\left[\frac{10}{s(s+10)}\right]$$

$$T(z) = \frac{z-1}{z} Z\left[\frac{1}{s} - \frac{1}{s+10}\right]$$

$$T(z) = \frac{z-1}{z}\left[\frac{z}{z-1} - \frac{z}{z-e^{-10T}}\right]$$

$$T(z) = \left[1 - \frac{z-1}{z-e^{-10T}}\right]$$

$$T(z) = \frac{1-e^{-10T}}{z-e^{-10T}}$$

$$T(z) = \frac{1-e^{-1}}{z-e^{-1}}$$

We have used PFE to get the terms inside the brackets to have the forms in Table 4.1, column 1.

Column 3 of Table 4.1 gives the corresponding z-transformed sampled signals. The result is given in the last equation, which could be used to find the digital filter gain curve or used to get the difference equation to actually code the digital filter.

Digital Signal Processing

7.4 The Bilinear Transform (BLT) Filter

One of the best digital IIR filter approximations to an analog filter is obtained using the **Bilinear Transform (BLT) method**. The formula for this method is not obtained in the same way as the formulas for the preceding two methods were, which was to mathematically force the outputs to be the same for a specific type of input at the sample times. Instead, the BLT formula uses the fact that for a Laplace transform $X(s)$ of a signal $x(t)$, its integral without initial conditions is given by $X(s)/s$. Thus multiplying the Laplace transform of a signal by $1/s$ is the same as integrating the signal in the time domain.

However, the following difference equation also approximately integrates the signal $x(t)$ by using its sampled values $x(nT) = x(n)$. Remember that the integral of a signal without initial conditions is just the area under the signal.

$$y(n) = y(n-1) + \frac{x(n) + x(n-1)}{2} * T$$

The preceding equation just says that the area under a signal after a new input sample $x(n)$ is just the previously computed area $y(n-1)$ plus the average of the new input and the previous input multiplied by the time between samples. Now this is just the previous area plus the average of the new area. By z-transforming the preceding difference equation, we get the following equations, which are solved for $T(z)$ once we get the transforms of the input $X(z)$ and the output $Y(z)$.

$$Y(z) = z^{-1}Y(z) + 0.5[X(z) + z^{-1}X(z)] * T$$

$$Y(z)(1 - z^{-1}) = 0.5 T X(z)(1 + z^{-1})$$

$$T(z) = \frac{Y(z)}{X(z)} = \frac{T}{2} \frac{1 + z^{-1}}{1 - z^{-1}}$$

The preceding equation is a transfer function of a DSP system that approximates the analog transfer function of a system that does integration. These two transforms are equated in the following form and then solved for s after multiplying numerator and denominator by z.

IIR Filter Design

$$\frac{T}{2}\frac{z+1}{z-1} \approx \frac{1}{s}$$

$$s = \frac{2}{T}\frac{z-1}{z+1}$$

This relationship is then used in any Laplace transfer function of an analog filter $T(s)$ to get the digital IIR filter $T(z)$ that approximates the analog filter. This is shown in Equation 7.3, which is the formula used to compute the BLT IIR digital filter, starting with the Laplace transfer function of an analog filter.

$$T(z) = T(s)|_{s = \frac{2}{T}\frac{(z-1)}{(z+1)}} \qquad \text{(Equation 7.3)}$$

Equation 7.3 says that the BLT digital IIR filter that approximates an analog filter can be obtained by using the equation for s to replace every s in the analog transfer function $T(s)$. This procedure is illustrated in Example 7.4.

Example 7.4. Using the BLT method to find the IIR filter for a first-order analog highpass filter

Problem: Given the following first-order highpass analog filter $T(s)$, use the BLT method to find an IIR digital filter $T(z)$ to approximate it for a sample time $T = 0.05$ s.

Solution:

0.05 $\qquad$ the value for T

$\dfrac{2}{T} \cdot \dfrac{z-1}{z+1}$ $\qquad$ the BLT substitution for the Laplace variable s

$\dfrac{3}{s + 10}$ $\qquad$ the first-order analog highpass filter

$\dfrac{2}{T} \cdot \dfrac{(z-1)}{\left[(z+1) \cdot \left[\dfrac{2 \cdot (z-1)}{T \cdot (z+1)} + 10\right]\right]}$ $\qquad$ using substitute for s

$\dfrac{(z-1)}{(z - 1 + 5 \cdot T \cdot z + 5 \cdot T)}$ $\qquad$ using simplify

Digital Signal Processing

$$\frac{(z-1)}{((1+5\cdot T)\cdot z - 1 + 5\cdot T)} \qquad \text{using collect on subexpression z}$$

$$\frac{(z-1)}{(1.25\cdot z - .75)} \qquad \text{using substitute for T}$$

$$4.\cdot \frac{(z-1.)}{(5.\cdot z - 3.)} \qquad \text{using simplify again}$$

Figure 7.7 shows the digital IIR filter gain in dB, while Figure 7.8 shows the gain in dB of the original analog filter.

Application 1

1. *Problem:* Design a digital lowpass IIR filter system to listen to or record the low-frequency sounds in water made by whales and other creatures that communicate over long distances underwater. The system should reduce the amplitudes of acoustic signals by less than 3 dB below 100 Hz, while reducing the amplitudes of acoustic signals above 300 Hz by more than 18 dB. These specifications should reduce the unwanted noise (due to wave action, water motion and turbulence, motor and hull, and electronic noise of the recording system) to an acceptable level. The digital filter system should include an appropriate anti-aliasing filter, and the design of the filter should result in a computer flow diagram, which could then be coded in C or any other high-level language.

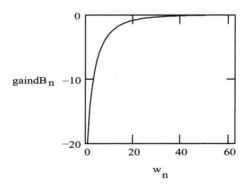

Figure 7.7. Gain plot of digital highpass filter of Example 7.4

IIR Filter Design

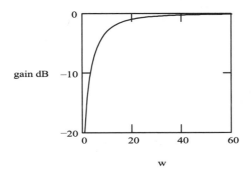

Figure 7.8. Gain plot in dB of highpass analog filter of Example 7.4

Solution: The preceding digital filter specifications are shown graphically in Figure 7.9, with an acceptable analog filter gain curve drawn in. Remember any gain curve in the clear areas is acceptable. The students can use their analog filter programs to determine an acceptable analog filter gain curve, or use their basic knowledge of frequency response plots, which is the approach taken next.

From Figure 7.9, it is seen that the corner frequency of the required filter is 100 Hz, and it must drop by 18 dB in 1.5 octaves. Since, from analog signal processing class, the students know that the gain curve drops approximately 6 dB per octave after the corner frequency per pole, then it is seen that the filter must be an analog filter with a constant numerator and a second-order denominator. A simple analog filter transfer function that does this is given by Equation 7.4.

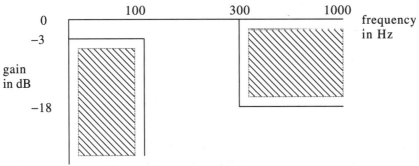

Figure 7.9. Lowpass filter graphical specification for Application 1

$$T(s) = \frac{(628)^2}{s^2 + 1.41(628)s + (628)^2} \qquad \text{(Equation 7.4)}$$

In Equation 7.4, the value of 628 is just the corner frequency in rad/s, and the 1.41 is just the damping ratio for no overshoot of a second-order system, as learned in analog signal processing. That these values give a filter with a gain curve that meets the graphical specifications can be easily seen by letting $s = jw$ in Equation 7.4, as shown here.

$$T(jw) = \frac{(628)^2}{-w^2 + 1.41(628)jw + (628)^2}$$

The magnitude of the preceding equation, which is the gain of the analog filter, shows that at very low frequencies ($w \ll 628$) the gain is approximately 1. However, at very high frequencies ($w \gg 628$), the gain is determined by the following equation, which drops by a factor of 4 for every increase in w by 2. This is just a drop of 12 dB per octave.

$$T(jw) \approx \frac{(628)^2}{-w^2}, \text{ for } w > 628$$

At $w = 628$ rad/s, the equation gives the gain as $1/1.41$, or -3 dB, as shown here.

$$T(jw) = \frac{(628)^2}{-(628)^2 + j1.41(628) + (628)^2} = \frac{1}{1.41j}$$

From the preceding discussion it is seen that an analog filter given by Equation 7.4 meets the graphical specifications of the lowpass filter. The next design step is to get an IIR digital filter to approximate the analog filter.

Using the BLT method shown in Equation 7.5, we can get an approximate digital IIR filter to replace the analog filter given in Equation 7.4. The sampling period T must be chosen before any digital filter can be derived, since we have learned that changing T changes the filter gain. For this application let's choose a sample frequency of 2 kHz, which gives $T = 0.0005$ s. Any sampling frequency substantially above twice the highest specified graphical frequency is appropriate. In this case the Nyquist limit is at 1 kHz, which is over an octave above the stopband frequency of 300

IIR Filter Design

Hz. The resulting digital IIR filter, $T(z)$, is given in Equation 7.6. The students should verify this for themselves on paper or using Mathcad.

$$T(z) = [T(s)]_{s = \frac{2}{T} \frac{(z-1)}{(z+1)}} \quad \text{(Equation 7.5)}$$

$$T(z) = \frac{0.01977(z+1)(z+1)}{z^2 - 1.565z + 0.6438} \quad \text{(Equation 7.6)}$$

Now that we have a possible IIR digital filter, its gain curve should be checked to see if our assumptions and computations are reasonable. Using the method given in Chapter 6 and shown in Equation 7.7, the gain is computed and plotted using Mathcad in Figures 7.10a and 7.10b.

$$\text{gain} = |T(z)|_{z = \cos(wT) + j\sin(wT)} \quad \text{(Equation 7.7)}$$

As can be seen in Figures 7.10a and 7.10b, the gain curve meets the graphical specifications except for the repeated passband at the sampling frequency. In Chapter 2 it was shown that this is unavoidable due to sampling; thus, we need to design an anti-aliasing filter that will reduce this repeated passband gain below the graphical specifications.

The anti-aliasing filter must reduce the gain of the system using it by at least 18 dB at 2 kHz, which is 12,566 rad/s. As a first-order lowpass analog filter drops 20 dB per decade above the corner frequency, the corner frequency should be at about 1,257 rad/s. Thus the transfer function of the analog filter is given by Equation 7.8.

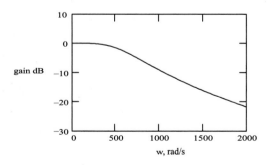

Figure 7.10a. Mathcad gain plot of IIR LPF

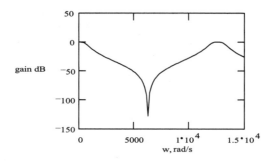

Figure 7.10b Mathcad gain plot of IIR LPF

$$T(s) = \frac{1257}{s + 1257} \quad \text{(Equation 7.8)}$$

The frequency response is given by letting $s = jw$, which gives the following equation.

$$T(jw) = \frac{1257}{jw + 1257}$$

It can be seen from the preceding equation that at low frequencies the gain (the magnitude of the frequency response) is 1, while at high frequencies it is reduced by about a factor of ten when the frequency is ten above the corner frequency. This is the same as saying a 20 dB reduction per decade.

The next task is to design an analog filter with the transfer function given in Equation 7.8. The RC circuit shown in Figure 7.11 has this transfer function, as shown next.

$$\frac{V_o}{V_i} = T(s) = \frac{\frac{1}{sC}}{R + \frac{1}{sC}} = \frac{1}{RCS + 1} = \frac{\frac{1}{RC}}{s + \frac{1}{RC}}$$

If C is chosen to be 0.1 µF, then R must be 8 K for the corner frequency of $1/RC$ to be about 1,257 rad/s. The frequency response of the analog anti-aliasing filter is obtained by letting $s = jw$ in Equation 7.8. The complete system is shown in Figure 7.12, and its total gain is obtained by

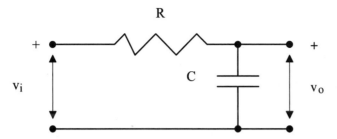

Figure 7.11 First-order RC analog anti-aliasing filter for Application 1

adding the gains of the anti-aliasing filter and the digital filter in dB. This is shown in Figure 7.13.

Finally, to implement the filter system, the RC lowpass anti-aliasing filter must be built. Its output must go into an ADC whose outputs are the sampled inputs to the processor running the digital filter. The coding for the digital filter is obtained by determining the difference equation from the digital filter transfer function and then coding the difference equation in a loop that saves the input and output values that are needed, while inputting new ADC samples and outputting new difference equation outputs to the DAC. First, the difference equation is derived, using the shifting property from z-transform theory, as follows.

$$T(z) = \frac{Y(z)}{X(z)} = \frac{0.01977(z+1)(z+1)}{z^2 - 1.565z + 0.6438}$$

$$Y(z)(z^2 - 1.565z + 0.6438) = X(z)(0.01977)(z^2 + 2z + 1)$$

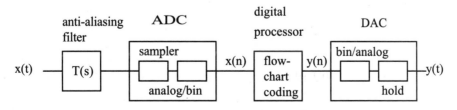

Figure 7.12. Complete system for Application 1

Digital Signal Processing

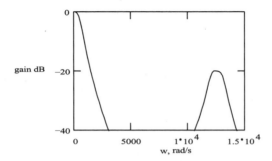

Figure 7.13 Mathcad gain plot of IIR LPF and analog anti-aliasing filter

$$z^2 Y(z) = 1.565 z Y(z) - 0.6438 Y(z) + 0.01977 z^2 X(z) + 0.03954 z X(z) + 0.01977 X(z)$$

$$Y(z) = 1.565 z^{-1} Y(z) - 0.6438 z^{-2} Y(z) + 0.01977 X(z) + 0.03954 z^{-1} X(z) + 0.01977 z^{-2} X(z)$$

$$y(n) = 1.565 y(n-1) - 0.6438 y(n-2) + 0.01977 x(n) + 0.03954 x(n-1) + 0.01977 x(n-2)$$

The flow diagram of the coding, which could be done in any of several languages like Basic or C, is shown graphically in Figure 7.14, where the programmer has assumed that the latest input $x(n)$ is given the name X, the previous input $X1$, and the input before that $X2$. The same coding scheme is used for the output $y(n)$, $y(n-1)$, and $y(n-2)$.

Summary

In this chapter we have finally developed formulas to determine several digital IIR filters to approximate analog filters. These formulas are given in Equations 7.1, 7.2, and 7.3 for the impulse invariant, step invariant, and BLT methods respectively. Mathematically the impulse invariant can be seen to be more direct, while the step invariant requires more computation to get $T(z)$. The BLT usually requires even more computation but is easy for a mathematical package to perform since it involves symbolic substitution of variables. As the BLT usually has a gain curve that more closely matches that of the analog filter, it is usually preferred. However, if the input is closely approximated by a steplike function, then

IIR Filter Design

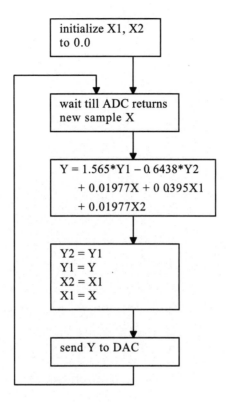

Figure 7.14. Flowchart for Application 1

the step invariant method will obviously be more accurate. In control systems the BLT method's phase shift may cause more problems than the step invariant, which is usually preferred for digital control. For digital filtering, the phase shift is not a problem.

In section 7.1 a review of analog filter approximations was done, since digital IIR filters are approximations to these analog filters. The digital IIR filter designer must start out with one of these analog approximations to the ideal analog gain specification.

Self-Test

1. For the following first-order lowpass analog filter $T(s)$, use the impulse invariant method to determine a digital IIR filter to replace it for the

given sample periods T. The analog filter has a corner frequency of 100 rad/s and a dc gain of 1.

$$T(s) = \frac{100}{s + 100}$$

(a) $T = 0.01$ s

(b) $T = 0.002$ s

2. For the first-order lowpass analog filter in Problem 1, use the step invariant method to determine a digital IIR filter to replace it for the given sample periods T.

(a) $T = 0.01$ s

(b) $T = 0.005$ s

3. For the following first-order highpass analog filter, use the BLT method to determine a digital IIR filter to replace it for the given sample periods T. The analog filter has a corner frequency of 50 rad/s and a highpass gain of 1.

$$T(s) = \frac{s}{s + 50}$$

(a) $T = 0.02$ s

(b) $T = 0.01$ s

4. For the following second-order Butterworth lowpass analog filter $T(s)$, use the impulse invariant method to determine the digital IIR filter to replace it for the given sample periods T. The analog filter has a corner frequency of 100 rad/s and a dc gain of 1.

$$T(s) = \frac{10000}{s^2 + 141.42s + 10000}$$

(a) $T = 0.01$ s

(b) $T = 0.005$ s

5. For the following second-order analog Chebyshev lowpass filter $T(s)$, use the impulse invariant method to determine the digital IIR filter to replace it for the given sample periods T. The analog filter has a corner frequency of 100 rad/s with a dc gain of 1 and a 3 dB ripple.

$$T(s) = \frac{7079}{s^2 + 64.5s + 7079}$$

(a) $T = 0.01$ s

(b) $T = 0.005$ s

6. For the second-order Butterworth lowpass analog filter $T(s)$ in Problem 4, determine the difference equation of the digital IIR filter that could then be coded on a computer or DSP chip for the sample period $T = 0.005$ s.

7. For the second-order Chebyshev lowpass analog filter $T(s)$ in Problem 5, determine the difference equation of the digital IIR filter that could then be coded on a computer or DSP chip for the sample period $T = 0.01$s.

8. For the highpass analog filter in Problem 3, determine the difference equations of the digital filter that could then be coded on a computer or DSP chip to replace the analog filter at the sample periods of $T = 0.02$ and $T = 0.01$ s.

9. For the first-order lowpass analog filter in Problem 1, use the BLT method to determine a digital IIR filter to replace it for the sample periods in Problem 1.

10. Use the method given in Chapter 6 and Mathcad to plot the gain in dB of the lowpass digital filters in Problem 1 for the two sampling periods given in Problem 1. Let the frequency axis go from 10 to 1,000 rad/s, using a log scale with a value computed every 10 rad/s.

11. Use the method given in Chapter 6 and Mathcad to plot the gain in dB of the lowpass digital filters given in Problem 2 for the two sampling periods given in Problem 2. Let the frequency be plotted on a log scale and go from 20 to 200 rad/s with values computed every 20 rad/s.

12. Use the method given in Chapter 6 and Mathcad to plot the gain in dB of the highpass digital filters given in Problem 3 for the two sampling periods given in Problem 3. Let the frequency axis be

plotted on a log scale and go from 10 to 200 rad/s with values computed every 10 rad/s.

13. For the following first-order highpass analog filter, use Mathcad to determine an IIR digital filter to replace it for the sample period of 0.025 s. Plot the gain in dB of the IIR filter from 1 to 100 rad/s on a log scale.

$$T(s) = \frac{2s}{(s+30)}$$

14. For the second-order Butterworth lowpass analog filter $T(s)$ given in Problem 4, use Mathcad to determine the digital IIR filter to replace it for the sample period $T = 0.008$ s.

15. For the second-order analog Chebyshev lowpass filter $T(s)$ given in Problem 5, use Mathcad to determine the digital IIR filter to replace it for the sample period $T = 0.003$ s. Plot the gain in dB from 10 to 1,000 rad/s on a log scale.

Problems

1. For the following first-order lowpass analog filter $T(s)$, use the impulse invariant method to determine a digital IIR filter to replace it for the given sample periods T. The analog filter has a corner frequency of 250 rad/s and a dc gain of 2.

$$T(s) = \frac{500}{s + 250}$$

   (a) $T = 0.004$ s

   (b) $T = 0.002$ s

2. For the first-order lowpass analog filter in Problem 1, use the step invariant method to determine a digital IIR filter to replace it for the given sample periods T.

   (a) $T = 0.004$ s

   (b) $T = 0.002$ s

3. For the following first-order highpass analog filter, use the BLT

method to determine a digital IIR filter to replace it for the given sample periods T. The analog filter has a corner frequency of 100 rad/s and a highpass gain of 3.

$$T(s) = \frac{3s}{s+100}$$

(a) $T = 0.005$ s

(b) $T = 0.0025$ s

4. For the following second-order Butterworth lowpass analog filter $T(s)$, use the impulse invariant method to determine the digital IIR filter to replace it for the given sample periods T. The analog filter has a corner frequency of 250 rad/s and a dc gain of 1.

$$T(s) = \frac{62500}{s^2 + 353.6s + 6250}$$

(a) $T = 0.005$ s

(b) $T = 0.001$ s

5. For the following second-order analog Chebyshev lowpass filter $T(s)$, use the impulse variant method to determine the digital IIR filter to replace it for the given sample periods T. The analog filter has a corner frequency of 250 rad/s and a dc gain of 1 with 3 dB ripple.

$$T(s) = \frac{44247}{s^2 + 161.2s + 44247}$$

(a) $T = 0.005$ s

(b) $T = 0.001$ s

6. For the second-order Butterworth lowpass analog filter $T(s)$ in Problem 4, determine the difference equation of the digital IIR filter that could then be coded on a computer or DSP chip for the sample period $T = 0.001$ s.

7. For the second-order Chebyshev lowpass filter $T(s)$ in Problem 5, determine the difference equation of the digital IIR filter that could then be coded on a computer or DSP chip for the sample period $T = 0.005$ s.

8. For the highpass analog filter in Problem 3, determine the difference

Digital Signal Processing

equations of the digital filter that could then be coded on a computer or DSP chip to replace the analog filter at the sample periods of $T = 0.005$ s and $T = 0.0025$ s.

9. For the first-order lowpass analog filter in Problem 1, use the BLT method to determine a digital IIR filter to replace it for the sample periods in Problem 1.

10. Use the method given in Chapter 6 and Mathcad to plot the gain in dB of the lowpass digital filters in Problem 1 for the two sampling periods given. Let the frequency axis go from 10 to 1,000 rad/s using a log scale with values every 25 rad/s.

11. Use the method given in Chapter 6 and Mathcad to log the gain in dB of the lowpass digital filters in Problem 2 for the two sampling periods given. Let the frequency axis be plotted in a log scale and go from 1 to 1,000 rad/s with values every 10 rad/s.

12. Use the method given in Chapter 6 and Mathcad to plot the gain in dB of the highpass digital filters given in Problem 3 for the two sampling periods given. Let the frequency axis be plotted in a log scale and go from 25 to 200 rad/s with values every 5 rad/s.

Answers to Self-Test

1a. $T(z) = \dfrac{z}{z - 0.368}$

1b. $T(z) = \dfrac{0.2z}{z - 0.817}$

2a. $T(z) = \dfrac{0.632}{z - 0.368}$

2b. $T(z) = \dfrac{0.393}{z - 0.607}$

3a. $T(z) = \dfrac{0.667(z - 1)}{z - 0.333}$

3b. $T(z) = \dfrac{0.8(z-1)}{z - 0.6}$

4a. $T(z) = \dfrac{0.460z}{z^2 - 0.75z + 0.243}$

4b. $T(z) = \dfrac{0.172z}{z^2 - 1.37z + 0.49}$

5a. $T(z) = \dfrac{0.463z}{z^2 - 1.03z + 0.525}$

5b. $T(z) = \dfrac{0.147z}{z^2 - 1.58z + 0.724}$

6. $y(n) = 1.32y(n-1) - 0.49y(n-2) + 0.172x(n-1)$

7. $y(n) = 1.03y(n-1) - 0.525y(n-2) + 0.147x(n-1)$

8. $y(n) = 0.333y(n-1) + 0.677[x(n) - x(n-1)]$,
   $y(n) = 0.6y(n) + 0.8[x(n) - x(n-1)]$

9a. $T(z) = \dfrac{0.333(z+1)}{z - 0.333}$

9b. $T(z) = \dfrac{0.0909(z+1)}{(z - 0.818)}$

13. gain of −23.5 dB at 1 rad/s, corner frequency at 30 rad/s

14. $T(z) = \dfrac{0.0927(z+1)(z+1)}{z^2 - 0.9735z + 0.3444}$

15. $T(z) = \dfrac{0.0143(z+1)(z+1)}{z^2 - 1.769z + 0.8261}$

chapter 8

Digital Filter and DSP Stability

Introduction

In Chapter 7 we learned to derive the mathematical description $T(z)$ of an IIR digital filter from the mathematical description $T(s)$ of an analog filter. Since IIR digital filters are approximations to corresponding analog filters, they have stability concerns just like any analog system given by $T(s)$ using Laplace transforms. For analog filters this is not much of a problem, since their design methods do not give unstable filters. However, if the filter is near instability, slight changes in the system could cause the output to grow without bound for any input. As we will see in Chapter 9, digital IIR filters are usually much more sensitive to numerical tolerances than their analog counterparts.

The consideration of stability will also give the student a deeper insight into how the mathematical description $T(z)$ of a digital filter or DSP system indicates many of the properties of the filter or system without having to code and test it. We show this by relating the stability of digital and analog systems to the pole locations of their transfer functions, and then relating the pole and zero locations of $T(z)$ to those of $T(s)$. The poles and zeros of $T(s)$ allow one to estimate the gain of an analog system using the familiar Bode plots.

8.1 Introduction to Stability

A system, including analog or digital filters, is **unstable** if its output grows without bound no matter what the input is, or even without an input but because of initial conditions or noise. For analog systems, if any root of the denominator of $T(s)$ is in the right half of the s-plane, it is unstable (as well as if there are repeated roots on the imaginary or jw axis). This is shown in Figure 8.1. The methods used to design analog filters do not generate unstable filters, but filters that are near being unstable have initial outputs that may be undesirable. The student will see in Chapter 9 that digital filters are usually much more sensitive to numerical tolerances than analog filters, as well as sensitive to changes in the sample period T for which it was designed. It would be nice to find a relationship between the denominator roots of $T(z)$ and digital filter stability, and even a relationship between the z-plane positions and s-plane positions.

An analog system is unstable if its transfer function $T(s)$ has any root of its denominator, called **poles**, in the **RHP** or right half of the s-plane. Most students are familiar with this from analog signal processing and control systems courses. The transfer function $T(s)$ is a ratio of polynomials, as is shown in the following equation of a second-order system.

$$T(s) = \frac{3s + 25}{s^2 + 4s + 35}$$

Even if the analog filter or system is of higher order, a fundamental property of algebra says that any order polynomial can be factored into products of first- and second-order polynomials with real coefficients. The roots of the denominator control some of the properties of the analog system; their location in the s-plane determines the stability or instability of the analog system. Since digital IIR filters or DSP systems approximate

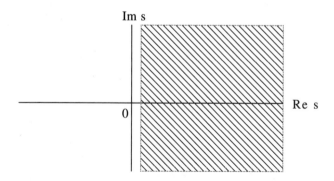

Figure 8.1 The hatched area is the unstable region for analog poles in s plane

the responses of analog systems, they could become unstable or nearly unstable. What is needed is a way to use the considerable knowledge about the pole locations in the s-plane to determine knowledge about the pole locations of $T(z)$ in the z-plane. This is done in the next section.

8.2 The Z-plane Unit Circle

In Chapter 4 we developed the z-transform of discrete-time signals. In Table 4.1 we give the z-transform $X(z)$ of a sampled exponential signal $x(n)$. This is given again in Equation 8.1, where $c = e^{-aT}$ is just a constant.

$$x(n) = A e^{-anT} = A c^n$$

$$X(z) = A\frac{z}{z - e^{-aT}} = A\frac{z}{z - c} \qquad \text{(Equation 8.1)}$$

In the development of the z-transform of the preceding discrete-time signal, there was no restriction put upon c. Usually c is real and positive, so that $x(n)$ is an exponentially decreasing sampled signal. However, it could be thought of as a zero, negative, imaginary, or even complex pole. By doing this, we will see that the poles of the z-transfer function $T(z)$ tell a lot about the stability and other behavior of the digital filter or DSP system described by $T(z)$.

The ability of poles of $T(z)$ to give us a qualitative idea about the output of a DSP system or digital filter is explained next. We have seen that $T(z)$ is the mathematical description of a DSP system, and if it is multiplied by the z-transform of an input we get the z-transform of the sampled output from the computer or DSP chip before it goes into the DAC to be converted back to an analog signal. The following equation shows this again.

$$Y(z) = T(z)R(z)$$

Now, a fundamental theorem of algebra—and the student's experience—says that the denominator polynomial of $Y(z)$ can be factored into products of first-order polynomials if you allow complex roots. Then using algebra to do a PFE, we could get $Y(z)$ to be a sum of terms with a constant in the numerator (times z) over a factor of the denominator polynomial of $Y(z)$. This is shown in the following equation without solving for the numerator coefficients, which may be complex.

$$Y(z) = \frac{b_2 z^2 + b_1 z + b_0}{(z-c_1)(z-c_2)(z-c_3)} R(z) = \frac{A_1 z}{z-c_1} + \frac{A_2 z}{z-c_2} + \frac{A_3 z}{z-c_3} + \cdots$$

The remainder of the terms are constants over the factors of the denominator of $R(z)$, the input z-transform. The inverse z-transform of the preceding equation could be taken by using Table 4.1 again to get the following equation.

$$y(n) = A_1(c_1)^n + A_2(c_2)^n + A_3(c_3)^n + \cdots$$

The preceding equation says that any output of a computer or DSP chip will be a sum of terms consisting of a constant multiplied by the pole of $T(z)$ raised to the nth power plus a term or terms due to poles of the input to the computer or DSP chip. The input is supplied by the user of the DSP system, but $T(z)$ is given by the DSP system designer, who should know what the terms of the output due to $T(z)$ are, no matter what the input is.

First let's look at a pole $c = 1$ in Equation 8.1. For this case we see that $x(n)$ has a term of constant amplitude A. This says that if any pole of $T(z)$ is +1, then the corresponding output sampled time $y(n)$ signal has a constant term in it. Notice that if the pole were slightly larger than +1, then the PFE of the output time signal would have a term in it that would grow without bound, which would give an unstable system no matter what the other terms were. The plot of this term for $c = 1.1$ and $A = 1$ is shown in Figure 8.2a.

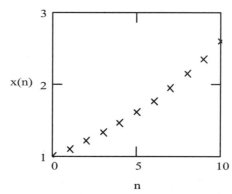

Figure 8.2a. Plot $x(n)$ of Equation 8.1 with pole c outside unit circle at 1.1

What about other pole positions? Again, using Equation 8.1 for poles of $-1, j, -j$, we have the following sampled time terms respectively, depending on the value of n.

$$A(-1)^n = +A, \text{ or } -A$$

$$A(j)^n = jA \text{ or } -A, \text{ or } -jA, \text{ or } A$$

$$A(-j)^n = jA \text{ or } -A, \text{ or } -jA, \text{ or } A$$

In the preceding equations the magnitude of the term with the pole of -1, j, or $-j$ is always 1. It can also be seen that if the magnitude of the poles were increased slightly, the corresponding PFE term would be an increasing magnitude in the sampled time domain. The opposite is true if the magnitudes of the poles were slightly less than 1. Thus it looks like if the magnitude of any pole of $T(z)$ is greater than 1, then the PFE term due to that term will have an increasing amplitude in the sampled time domain. This is another way of saying that the digital filter or DSP system is unstable if any pole lies outside a unit circle, as shown in Figure 8.2b.

Now remember this is just what happens with poles of the analog transfer functions to the right of the jw or imaginary axis in the s-plane. Also, analog poles on the jw axis for $w = 0$ give a constant term in the PFE just like the z-transfer function pole at $z = 1$. Thus, starting at $z = 1$ and traveling around the unit circle in the z-plane is similar to starting at $w = 0$ on the jw axis and moving up or down, except now you travel in a circle, which means you start to repeat. This corresponds to what we discovered

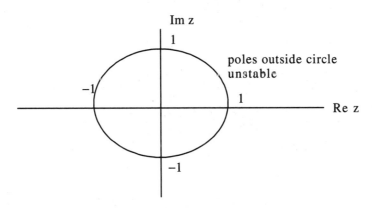

Figure 8.2b. Unit circle in the z-plane

in Chapter 2, that the frequency responses of digital filters or DSP systems repeat!

Thus, the line of the unit circle leaving from 1 in the z-plane is just like going up the jw axis starting at the origin in the s-plane with the stable region to the left and the unstable region for poles to the right. For the z-plane, right of the line is the region outside the unit circle, and left of the line is inside the unit circle. You can think of the jw axis as bending to the left till it closes on itself and the origin moved to 1 in order to form the stable region of the z-plane from the stable region of the s-plane. This again is illustrated in Figure 8.2b. In reverse order, you can think of the unit circle being cut at the −1 point and unwrapped into two vertical lines moved left to 0 to form the stable region of the s-plane, as shown in Figure 8.3. These visualizations give insight into the characteristics of the digital filter or DSP system given by $T(z)$ by thinking of the unwrapped unit circle with the inside area as going into the entire left half plane of the s-plane.

If you have a pole of $T(z)$ at 1, it is like an s-plane pole at 0. Any pole of $T(z)$ on the real axis inside the unit circle is like a pole of $T(s)$ on the negative real axis. Any pole of $T(z)$ inside the unit circle but off the real axis is like complex poles of $T(s)$ in the left half plane. Figure 8.4 illustrates these relationships. These analogies give insight into the characteristics of the digital filter $T(z)$. Examples 8.1 and 8.2 show how to use this relationship to determine the stability of a digital system.

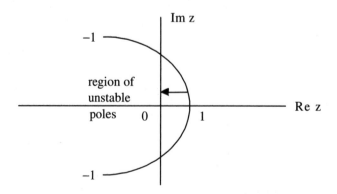

Figure 8.3. The unit circle becoming the *s* plane *jw* axis

Digital Filter and DSP Stability

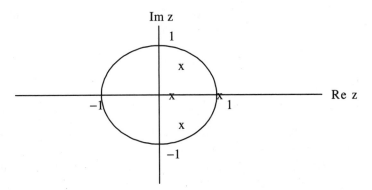

Figure 8.4a. Real poles and complex poles in z plane

Example 8.1. Using z-plane pole locations to determine if an IIR filter is unstable

Problem: Given the following digital IIR filter, determine if it is unstable, that is, its output will grow without bound for any input.

$$T(z) = \frac{z^2 + z + 1}{(z - 0.9)(z + 1.1)}$$

Solution: Without knowing how this filter was developed and what it was for, the student can still tell that this is an unstable filter as a result of a

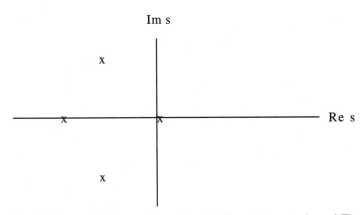

Figure 8.4b. S plane poles corresponding to the z-plane poles of Figure 8.4a

design error or numerical tolerances in determining the filter coefficients. The poles and zeros are given here.

$$\text{poles: } 0.9, -1.1$$

$$\text{zeros: } -0.5 \pm 0.87j$$

When determining stability, the student need only to look at the magnitude of the poles; as one is greater than 1, the digital filter is unstable. Note the instability is not caused by the positive pole at 0.9, which would cause instability in an analog system. Figure 8.5 shows the gain plot of the digital filter for $T = 0.1$ s, using the methods given in Chapter 6. This plot gives no hint of instability, but if the output of the digital filter were computed for any sinusoidal input, part of the output would be growing without bound. This part is shown in Figure 8.6.

Example 8.2. Using pole location to determine the stability of an unfactored $T(z)$

Problem: Let the digital filter transfer function $T(z)$ be given by the following equation.

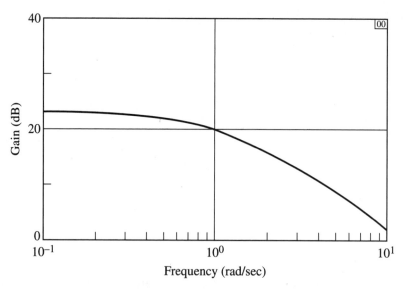

Figure 8.5. Gain plot of $T(z)$ in Example 8.1

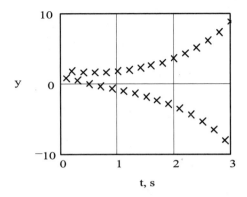

Figure 8.6. Time response for filter in Example 8.1

$$T(z) = \frac{0.2z^2 + 0.5z - 1}{z^2 - 1.6z + 1.28}$$

Solution: By using a calculator or completing the square, we can find the pole values.

$$z^2 - 1.6z + 1.28 = z^2 - 1.6z + \left(\frac{1.6}{2}\right)^2 - \left(\frac{1.6}{2}\right)^2 + 1.28$$

$$= (z - 0.8)^2 + 0.64$$

By setting the preceding equation to zero, we get the poles or roots of the denominator as:

$$\text{poles: } 0.8 + j0.8, \, 0.8 - j0.8$$

The magnitude of either pole is just the square root of the square of the real part added to the square of the imaginary part; either pole has a magnitude as calculated by the following equation.

$$\text{magnitude} = \sqrt{(0.8)^2 + (0.8)^2}$$

$$\text{magnitude} = \sqrt{1.28} = 1.13$$

Because the magnitude of either pole is greater than one, it is apparent that the digital filter is unstable and an error in design or tolerances was made in the coefficients. This example also illustrates a mathematical shortcut. If the quadratic factor of the denominator of $T(z)$ has complex roots with the coefficient of the z^2 term being one, then the magnitude of either complex root is the square root of the constant term of the quadratic factor.

8.3 Other Properties Using the Z-plane

Another way to arrive at the same result for stability, using pole location, but to get even more quantitative results is to use the relationship given by Equation 6.3. This equation is repeated here.

$$z = e^{sT} \qquad \text{(Equation 6.3)}$$

If we let $s = jw$, then this is just the jw axis in the s-plane, but this gives a value of z with magnitude of 1, as shown in the following equation after comparing the results with the Euler equation.

$$z = e^{jwT}$$

This is just a complex number of unit magnitude and angle of wT. As w goes up the jw axis in the s-plane, we can see that in the z-plane the value of z moves counterclockwise around the unit circle, since the angle wT is increasing. If s were to the left of the jw axis by a distance a, then we would have $s = -a + jw$, which gives the following equation for z.

$$z = e^{-aT} e^{jwT}$$

This is just a smaller circle inside the unit circle as w varies up and down the vertical line $-a$ to the left of the jw axis in the s-plane. This is shown in Figure 8.7 for $c = e^{-aT}$.

This relationship is very significant. It says that a pole of $T(z)$ in the z-plane at an angle of wT above the real z axis is like a position at w above the real axis in the s-plane. The magnitude of the pole of $T(z)$, $e^{-aT} = c$, in the z-plane away from the origin, is like a pole at a distance a into the left half s-plane. This is also true for a zero. Putting these two statements together, we have the following equations, which allow the poles and zeros of $T(z)$ to be thought of as poles and zeros of an analog system $T(s)$. This relationship is also shown in Figure 8.8.

$$z = e^{-aT} e^{jwT} = |z| \angle z \rightarrow s = -a + jw$$

Digital Filter and DSP Stability

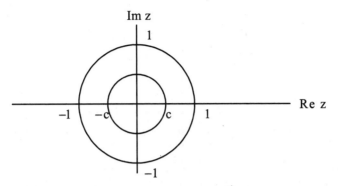

Figure 8.7. Unit circle with circle of radius $c = e^{-at}$, where $-a$ is the real part of s plane pole

Or in a more useful form, we have Equation 8.2, where z is a pole or zero location.

$$z = |z| \angle z \rightarrow s = (\ln|z| + j\angle z)/T \qquad \text{(Equation 8.2)}$$

Remember, the magnitude of a pole or zero of z is just its distance from the origin, and its angle is the angle up from the real z axis.

The preceding equation agrees with the qualitative results obtained earlier based on stability. If the magnitude of the pole at a value of z is less than 1, then it is in the left half of the s-plane, since the natural log

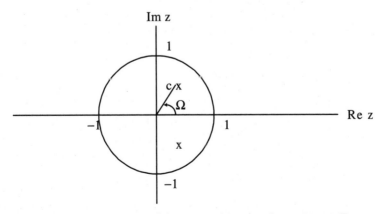

Figure 8.8a. Poles in z plane at $|z| = c$, and angle of $z = \Omega$ at wT

Digital Signal Processing

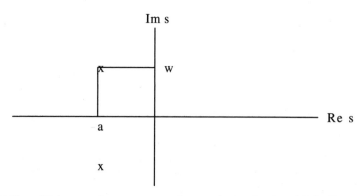

Figure 8.8b. S plane poles corresponding to the z plane poles in Figure 8.8a

of a number less than 1 is negative; and if the magnitude is greater than 1, it is in the right half of the s-plane. But this equation can be used for more quantitative evaluation of the effects of the poles and zeros of $T(z)$ by computing the corresponding s-plane locations. Remember, by knowing the poles and zeros (roots of the denominator or numerator) of $T(s)$, the Bode plot or gain plot could be approximated for the analog filter. Thus by using the relationship given in Equation 8.2, we can approximate the characteristics of the DSP system given by $T(z)$, and even determine the effects of changes to the system. Example 8.3 illustrates the use of Equation 8.2 to estimate the type of digital filter from $T(z)$, and even to estimate its critical frequencies if it is stable.

Example 8.3. Using the unit circle to determine the characteristics of a digital filter

Problem: Let $T(z)$ be as given here, determine if it is stable and, if so, the filter type and corner frequency.

$$T(z) = \frac{z^2 + z + 0.5}{z^2 - 2.00z + 0.99}$$

Solution: From the earlier discussions, we know that stability is determined only by the poles of $T(z)$. Factoring the denominator only, in order to see the roots or poles of $T(z)$, we get the following equation.

$$T(z) = \frac{z^2 + z + 0.5}{(z - 0.900)(z - 1.100)}$$

From the preceding equation we can see that there is a pole at 1.1, which is greater than 1, so the DSP system given by $T(z)$ is unstable; that is, its output will grow without bound due to an input or initial conditions or noise.

In the next example we will use the pole and zero locations of $T(z)$ with respect to the unit circle to determine the relative positions of poles and zeros in the s-plane that could be used to estimate the gain plot.

Example 8.4. Using the relationship of the z-plane to the s-plane to determine stability, type of filter, and filter corner frequency

Problem: Let a digital IIR filter transfer function $T(z)$ be given by the following equation with the sampling period $T = 0.1$ s. Use the relationships given in Equation 8.2 to determine the filter stability, filter type, and corner frequency.

$$T(z) = \frac{0.333(z + 1)}{z - 0.333}$$

Solution: The pole and zero of the digital filter in the z-plane are given here.

$$\text{pole: } z = 0.333$$

$$\text{zero: } z = -1$$

As the magnitude of the only pole is 0.333, the digital filter is stable and the gain plot really gives the gain of the filter. Using Equation 8.2 on the pole and zero we get the equivalent s-plane poles.

$$\text{pole: } s = 10\ln(0.333) + j0$$

$$= -11 + j0$$

$$\text{zero: } s = 10\ln(1) + j10(3.14)$$

$$= j31.4$$

Thus the corner frequency for the pole is 11 rad/s. The "corner frequency" at 31.4 rad/s is not an actual corner frequency; it is where the filter goes to zero at the aliasing frequency. Thus the digital filter is a lowpass filter with a corner frequency at 11 rad/s. Figure 8.9 illustrates the actual gain plot of the digital filter.

Summary

In this chapter the stability of digital filters and DSP systems was shown to be related to the pole locations of $T(z)$. Any pole outside the unit circle in the z-plane will cause the output of the DSP system to grow without bounds for any input. This corresponds to the unstable region of analog systems being the righthand plane of the s-plane. This led to a useful correlation between the z-plane and the s-plane pole locations. A pole at 1 in the z-plane is like a pole at 0 in the s-plane, and poles between 1 and 0 in the z-plane are like poles on the negative real axis in the s-plane.

By using the relationship given by Equation 6.3 between z and s, more quantitative results than the preceding ones were developed. In fact, Equation 8.2 showed how to convert a z-plane pole or zero position to an

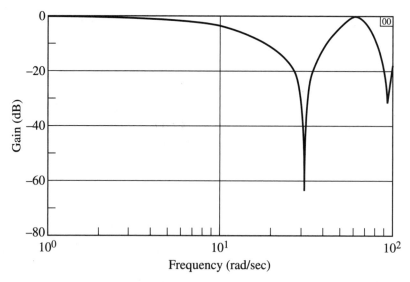

Figure 8.9. Gain plot of $T(z)$ for Example 8.4

Digital Filter and DSP Stability

approximately equivalent s-plane position. With this relationship the equivalent corner frequencies can be obtained as if the DSP system or filter were an analog system or filter. Then, using Bode plots, the gains of the digital filter or DSP system can be drawn. Examples were given showing this powerful method of visualizing the gain of digital filters by looking at the poles and zeros of $T(z)$.

Self-Test

1. Determine if the following digital filter is stable or unstable, and state your reason in terms of the pole magnitude.

$$T(z) = \frac{20z}{z+1.1}$$

2. Determine if the following digital filter is stable or unstable, and if unstable state which pole or poles make it unstable.

$$T(z) = \frac{20(z+1.5)}{(z+0.8)(z+0.5-0.9j)(z+0.5+0.9j)}$$

3. Determine if the following system is stable; if it is, determine if it is a highpass, lowpass, bandpass, or bandstop filter.

$$T(z) = \frac{0.8(z-1)}{z-0.6}$$

4. Determine if the following system is stable; if it is, determine if it is a highpass, lowpass, bandpass, or bandstop filter.

$$T(z) = \frac{0.2(z+1)}{z-0.6}$$

5. Determine the analog corner frequency of the pole in Problem 3 for $T = 0.1$ s.

6. Determine if the following digital filter is stable or unstable; if it is unstable, state the magnitude of the pole that makes it unstable.

$$T(z) = \frac{10z+2}{(z-0.2)(z-0.5)(z^2+1.2z+2)}$$

7. Determine the zero and pole values for an analog filter corresponding to the digital filter in Problem 3 for $T = 0.1$ s.

8. Determine the zero and pole values for an analog filter corresponding to the following digital filter below for $T = 0.1$ s.

$$T(z) = \frac{5(z-1)}{z - 0.368}$$

9. Determine the zero and pole values of an analog filter corresponding to the following digital filter below, and state if it is a highpass or lowpass filter when the sampling period $T = 0.03$ s.

$$T(z) = \frac{0.2(z - 0.970)}{z - 0.861}$$

Problems

1. Determine if the following digital filter is stable or unstable, and state your reason in terms of the pole magnitude.

$$T(z) = \frac{0.981z}{z - 0.782}$$

2. Determine if the following digital filter is stable or unstable; if unstable, state which pole or poles make it unstable.

$$T(z) = \frac{10(z - 0.90)z^2}{(z - 1.1)(z^2 - 1.20z + 0.95)}$$

3. Determine if the following system is stable, and if it is, determine if it is a highpass, lowpass, bandpass, or bandstop filter.

$$T(z) = \frac{1.2(z - 0.2)}{z - 0.6}$$

4. Determine if the following system is stable, and if it is, determine if it is a highpass, lowpass, bandpass, or bandstop filter.

Digital Filter and DSP Stability

$$T(z) = \frac{0.5(z-1)}{z-0.8}$$

5. Determine the analog corner frequency of the pole in Problem 3 for $T = 0.05$s.

6. Determine if the following digital filter is stable or unstable; if it is unstable, state the magnitude of the pole that makes it unstable.

$$T(z) = \frac{25z^2}{(z-1.1)(z-0.9)(z^2+1.4z+0.5)}$$

7. Determine the zero and pole values for an analog filter corresponding to the digital filter in Problem 3 for $T = 0.02$ s.

8. Determine the zero and pole values for an analog filter corresponding to the following digital filter for $T = 0.002$ s.

$$T(z) = \frac{4(z-1)}{z-0.5}$$

9. Determine the zero and pole values of an analog filter corresponding to the following digital filter, and state if it is a highpass or lowpass filter when the sampling period $T = 0.003$ s.

$$T(z) = \frac{0.89(z-0.8)}{z-0.5}$$

Answers to Self-Test

1. Unstable. The magnitude of the pole is greater than 1.

2. Unstable. The poles at $-0.5 \pm j0.9$ make it unstable.

3. Stable. It is a highpass filter.

4. Stable. It is a lowpass filter.

5. 5 rad/s

6. 1.414

7. 0, −5

8. 0, −10

9. −1, −5, highpass

chapter 9

Filter Coefficient Precision

Introduction

In Chapter 8 we learned to find the corner frequencies of an analog filter that would have approximately the same gain characteristics as a digital filter until it began to repeat. This allows the designer to see if the digital filter or DSP system is stable, and also to determine the type of filter and even the shape of the gain curve using Bode plots. In this chapter we look at the effect of small changes to the filter or system coefficients, the a and b coefficients of the difference equation. We see a surprising result, that a very small percentage change in these coefficients may cause a larger percentage change in the corner frequency of the corresponding analog filter. The exact change will be found to be a function of the sampling period T.

9.1 Introduction into Computer Numeric Precision

In analog filtering, part of the accuracy of the design is determined by the precision of the resistors and capacitors used in building the filter. In a similar way the accuracy of IIR digital filters and systems depends on the precision of the numbers used to represent the a and b coefficients of the filter difference equation to be coded. In digital filtering there is one more parameter, the sampling period T. We will find that as the sampling rate increases, or T decreases, more precision is required.

Precision is only a part of the accuracy of the coefficients. **Precision** is how well the value of the coefficient is represented in the coding regardless if the coefficient value is correct or not. For any binary numbering system used in a computer (usually two's complement is used), the accuracy of representation of the value of the coefficient, or precision, doubles for every extra bit used to code the number. (The student should be familiar with this from digital classes.) Thus the number of bits used to represent coefficients in the coding is analogous to the percentage tolerances used for component values for analog systems. But we will find that to design an IIR digital filter to the same accuracy as an analog filter, we may need 0.1 percent tolerance on the digital coefficients, but only 5 percent tolerances on the analog components. Example 9.1 shows how to compute the precision of a binary representation of an analog value.

Example 9.1. Determining the precision of a digital representation of a number

Problem: Let it be required to represent the b coefficient of 0.9 with a precision of 5 percent as a binary number.

Solution: Let's first try to use only two bits to the right of the binary point, and find the error. Using two bits, the binary representation is (0.)11, which is 0.75 base 10. Thus there is an error of almost 17 percent. Next, let's use three bits. The best we can do with three bits is (0.)111, which is 0.875 base 10. This gives an error of about 3 percent. Thus we see that three bits are required to represent a coefficient of 0.9 to less than 5 percent accuracy. If more bits were used, we could represent the coefficient even more precisely.

In Example 9.1, if the desired value were 0.75, then two bits would have been sufficient, but since many different values are required, enough bits are used to represent all the coefficient values to the required precision. The precision or number of bits gives the accuracy of the representation of the numeric value of the coefficient, but says nothing about how accurate the original numeric value was. Thus we say the "precision of the coefficient" when referring to the number of bits, and not its total accuracy.

The more precise a coefficient needs to be, the more bits to the right of the binary point are required. Each added bit to the right becomes the

Filter Coefficient Precision

Least Significant Bit, or LSB, and adds half the value of the previous bit. The maximum error by using n bits to the right is half the value of the LSB, since this is the maximum difference between the coefficient with and without the nth bit as the LSB. Using this we can determine the number of bits to the right of the binary point that are needed. This is shown in Example 9.2.

Example 9.2. Determining the number of bits to ensure a specified percentage precision

Problem: Find the number of bits required to represent the number 5.3 to 1 percent accuracy.

Solution: To be within 1 percent of 5.3, the binary number must be within 0.053 of 5.3. This requires n to be 4, since $1/2^n$ is 0.0625, and half of that is approximately 0.03.

9.2 Development of Equations for Precision Effects

Let us first look at deriving IIR digital filters to approximate single-pole analog filters, and then derive the difference equations for the IIR filters that would be coded. This will show that a percentage change in the corner frequency of an analog filter pole may require the percentage change of the b coefficient of the corresponding difference equations to be even less. Remember from analog filtering that given the dc gain and the corner frequencies, good straight-line approximations to the gain in dB of the frequency response may be obtained. These are called Bode magnitude plots. We will do this in Example 9.3 for a first-order lowpass analog filter. Example 9.3 develops the IIR filters for analog filters with two different corner frequencies, and then compares the change in the b coefficients of the corresponding difference equations of each IIR filter. This will illustrate that small changes in the b coefficients may correspond to larger percentage change in the pole corner frequencies.

Example 9.3. Computing the BLT IIR filters for first-order lowpass analog filters

Problem: Given the following two similar analog filters, determine the corresponding IIR filters using the BLT method from Chapter 7. The student can use the method to compute the results obtained here. The sampling period is given as $T = 0.01$ s. Then obtain the corresponding

IIR filter difference equations and compare the change in the b coefficient to the change in the analog filter corner frequencies.

$$T_1(s) = \frac{10}{s+10}$$

$$T_2(s) = \frac{20}{s+20}$$

Solution: Using the BLT method, we get the following IIR digital filters that approximate the preceding analog filters.

$$T_1(z) = \frac{0.04762(z+1)}{z-0.90476}$$

$$T_2(z) = \frac{0.0909(z+1)}{z-0.81818}$$

The corresponding difference equations are obtained by using the definition of $T(z)$, cross multiplying, and then using the shifting property, as the following equations show.

$$T_1(z) = \frac{Y_1(z)}{X_1(z)} = \frac{0.04762(z+1)}{z-0.90476}$$

$$T_2(z) = \frac{Y_2(z)}{X_2(z)} = \frac{0.0909(z+1)}{z-0.81818}$$

$$Y_1(z)(z-0.90476) = 0.04762(z+1)X_1(z)$$

$$Y_2(z)(z-0.81818) = 0.0909(z+1)X_2(z)$$

$$zY_1(z) - 0.90476Y_1(z) = 0.047619zX_1(z) + 0.047619X_1(z)$$

$$zY_2(z) - 0.81818Y_2(z) = 0.0909zX_2(z) + 0.0909X_2(z)$$

$$Y_1(z) - 0.90476z^{-1}Y_1(z) = 0.047619X_1(z) + 0.047619z^{-1}X_1(z)$$

$$Y_2(z) - 0.81818z^{-1}Y_2(z) = 0.0909X_2(z) + 0.0909z^{-1}X_2(z)$$

Filter Coefficient Precision

$$y_1(n) = 0.90476 y_1(n-1) + 0.047619 x_1(n) + 0.047619 x_1(n-1)$$

$$y_2(n) = 0.81818 z^{-1} y_2(n) + 0.0909 x_2(n) + 0.0909 x_2(n-1)$$

Notice in the preceding difference equations corresponding to two similar analog filters, with one having a corner frequency 100 percent greater than the other, that the percentage change in the b coefficients of the difference equations is only about 0.1 or 10 percent. However, the change in the analog filter corner frequencies was 100 percent.

Example 9.3 gives an interesting result that suggests further study of b coefficient accuracy is warranted. We found that a 10 percent change in one of the coefficients (the b coefficient) of the difference equation for a digital IIR filter corresponded to a 100 percent change in the analog filter corner frequency. We need to be able to predict how much precision is needed in the b coefficients for a specified corner frequency accuracy of the original analog filter being replaced by a digital IIR filter.

Let's look at the relationship between the poles of the Laplace transform of the following exponentially decaying signal $f(t)$, and poles of the z-transform of the sampled signal. These signals are given in the following equations.

$$f(t) = A e^{-at}$$

$$F(s) = \frac{A}{s+a}$$

$$F(z) = \frac{Az}{z-c}$$

In the last equation, $c = e^{-aT}$, where T is again the sampling period. Thus a Laplace transform pole at $-a$ corresponds to a z-transform pole at $c = e^{-aT}$. We will use this relationship of poles of signals to find the relationship between poles of the Laplace transfer function $T(s)$ of the analog filter and the z-transfer function $T(z)$ of an equivalent IIR digital filter.

Let's first look at what the Laplace transform pole at $-a$ means. If s is replaced by jw in the preceding Laplace transform of the exponentially decaying signal starting at A, we have the following equation.

$$T(jw) = \frac{A}{jw + a}$$

This is just the frequency response equation of the analog system $T(s)$. We can see that for the frequency w below a rad/s, the dc gain is A/a. Above the frequency of a rad/s the gain plot drops by a factor of ten for every increase in frequency of ten, or in dB it drops 20 dB/decade. This is the straight line Bode plot approximation of the gain in dB with respect to frequency. This plot is shown in Figure 9.1. The straight-line approximation has its greatest error at the corner frequency a, where the gain in dB is down 3 dB from A/a in dB. If the pole is one of a pair of complex poles, the maximum error is still near the corner frequency, but the error value depends on the damping ratio of the complex pole pair.

Let's look at what the z-plane pole at c means. If we use the fact that $T(z)$ is the ratio of $Y(z)$ over $R(z)$, and cross-multiply and use the shifting property to get the difference equation that is coded, we get the following equations.

$$T(z) = \frac{Y(z)}{X(z)} = \frac{Az}{z - c}$$

$$Y(z)(z - c) = X(z)Az$$

$$zY(z) - cY(z) = AzX(z)$$

$$Y(z) - cz^{-1}Y(z) = AX(z)$$

$$y(n) - cy(n-1)y(n) = Ax(n)$$

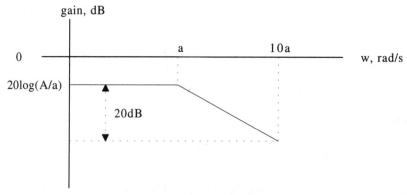

Figure 9.1 Straight-line Bode plot for gain in dB of single-pole filter

From the last equation, we can see that c is the b_1 coefficient and A is the a_0 of the modified difference equation. Thus, for a first-order system or filter (one with only one pole), we can finally see that a is the corner frequency of the analog system and c is the only b coefficient.

Now that we know that a pole at $-a$ gives the corner frequency of the analog system at a, and that a pole at c gives the b coefficient value of the digital filter difference equation, we would like to be able to predict the effect of a change in the b coefficient on the desired corner frequency. We can do this by using the relationship given earlier between an analog pole at $-a$ and a z-plane pole at c. This relationship is repeated as Equation 9.1.

$$c = e^{-aT} \qquad \text{(Equation 9.1)}$$

By taking the natural log of both sides of Equation 9.1, we get Equation 9.2, which is in a more useful form.

$$a = -\frac{\ln(c)}{T} \qquad \text{(Equation 9.2)}$$

This equation says that the smaller the sampling period T, the larger is the change in a for a given change in c. Example 9.4 uses Equation 9.2 to predict the results found in Example 9.3, but in a much easier way.

Example 9.4. Predicting the change in the analog corner frequency due to a change in the difference equation b coefficient

Problem: Given the difference equation for $y_1(n)$ in Example 9.3 for an IIR digital filter with a corner frequency at 10 rad/s, determine the percentage change in the analog filter corner frequency if the b coefficient of the difference equation is changed by 10 percent.

Solution: Using Equation 9.2, where c is b_1 for the original digital filter and then b_1 for the changed digital filter, and T is 0.01 s, we get the following equations.

$$a = \frac{-\ln(0.9)}{0.01} = 10.5$$

$$a = \frac{-\ln(0.8)}{0.01} = 22.3$$

We can see that even using approximate values for the b coefficient, a 10 percent change from 0.9 to 0.8 causes an almost 100 percent change from 10.5 to 22.3 in the corner frequency.

In Example 9.4 we verified the results of Example 9.3 in just a few short steps using Equation 9.2, where the pole at c represented the b coefficient of the IIR digital filter. Notice that in Example 9.4 the answer seems to depend upon the value of the sampling period T. This is shown by Equation 9.2, where as T decreases the change in a, the change in the corner frequency of the filter represented by the digital filter, increases for the same change in b. This is true even though in Chapter 7 we learned that as T changes, $T(z)$ and thus the b value also change. This is illustrated in Example 9.5.

Example 9.5. Illustration of the effect of the sampling period on required coefficient precision

Problem: Repeat Example 9.4 for $T = 0.001$ s.

Solution: The b_1 coefficient (the only b coefficient) for the difference equation for the first analog filter using the BLT method is now approximately 0.99, and a reduction of 10 percent approximately gives 0.89 for the new b. Using Equation 9.2 we have the following equations.

$$a = \frac{\ln(.99)}{0.001} = 10$$

$$a = \frac{-\ln(0.89)}{0.001} = 116$$

In Example 9.5 we can see that, at least for first-order digital IIR filters, as the sampling period T is reduced, we get even more sensitivity of the analog corner frequency accuracy to small percentage changes in the b coefficient of the difference equation for the corresponding digital IIR filter. This statement is true qualitatively for higher-order digital IIR filters.

We have been looking at the property of a digital IIR filter with one pole. What about digital filters with two or more poles and the number of bits required to ensure a specified corner frequency accuracy? The effect of changes due to a lack of precision in the b coefficient of one pole then becomes more of a qualitative statement about the precision required for the b coefficients of filters with more than one pole. It can usually predict the number of bits required to ensure the corner frequency accuracies within an order of magnitude. This is just the number of decimal digits required in the coding of the difference equation. If the digital IIR filter is sampled at a frequency much higher than the corner frequencies of the analog filter, then the precision required to ensure the accuracy of the only b coefficient of a one-pole digital IIR filter corresponding to the lowest corner frequency is a very good estimate of the precision of all the b coefficients of digital filters with more poles (if all the poles have the same accuracy specification). This is shown in Example 9.6.

Example 9.6. Determining the precision of the b coefficients of a two-pole digital IIR filter

Problem: Given the following bandpass filter $T(s)$ with corner frequencies at 10 and 40 rad/s, estimate the number of decimal digits to the right of the decimal point that are needed to ensure that the IIR digital filter obtained using the BLT method approximates the corner frequencies of the analog filter within 10 percent for $T = 0.001$.

$$T(s) = \frac{40s}{(s+10)(s+40)}$$

Solution: The lowest corner frequency of the analog filter is 10 rad/s. If we wanted to use this as the only pole of the analog filter and find the only b coefficient using the BLT method, we would get $b = c = 0.99005$. Trying a 1 percent reduction in b, or $b = 0.98$, gives $a = 20$, which is a 100 percent change in a. Next try a 0.1 percent change in b. This gives $a = 11$, or a 10 percent change. Thus we will need to use three decimal digits to the right of the decimal point to ensure a 10 percent accuracy in the corner frequencies represented by the digital IIR filter.

In this section we used Example 9.3 to show that there is indeed more sensitivity to the percentage change of a b coefficient than the percentage change specified for the corner frequency of the corresponding analog single-pole filter. Then we used the relationship between the analog and digital poles for a decaying exponential signal to show this is true for all single-pole systems. We also used the relationship between the analog and digital poles to derive Equation 9.2, which gives the change in a, the corner frequency, given a change in c, the b coefficient. This approach was used to give confidence to the results by showing the problem and how the poles are related and what they represent in the analog and digital domains. However, a shorter method could have been used by using the purely mathematical relationship given in Table 4.1 for c, which is e^{-aT}. If the natural log is taken of both terms, noting that $c = b$, then we immediately get Equation 9.2.

The next question to answer is: Does Equation 9.2 give the change in the corner frequency of an analog filter with a single zero if the c in the equation is replaced by an a coefficient rather than a b coefficient? It turns out this is true. This is true since Equation 6.3 says $z = e^{sT}$. For a real pole or zero, $s = -a$, where a is positive. This gives an important relationship that was given in Table 4.1 for a pole, but can be seen from the math to apply to poles and zeros. This relationship is shown in Equation 9.3.

$$z = e^{-aT} \qquad \text{(Equation 9.3)}$$

This relationship is just a mapping between the s-plane and the z-plane given the sampling period T. We already know that the LHP of the s-plane goes into the inside of the unit circle of the z-plane. This relationship gives a more precise relationship between poles and zeros. If there is a pole or zero at $-a$ in the s-plane, the z-plane will have a corresponding pole or zero at the value computed by using Equation 9.3. This is the same relationship that lead to Equation 9.2, and thus this equation can be used for both poles and zeros of analog filters. This is shown using a single-zero analog LPF in Example 9.7.

Example 9.7. Determining the change in the analog zero corner frequency due to a change in the a coefficient

Problem: Given the following analog lowpass filter $T(s)$ with a single zero with the sampling period $T = 0.01$. Determine what the percentage change in the corner frequency of the analog filter would be for a 4 percent reduction in the a coefficient of the corresponding IIR filter difference equation.

$$T(s) = \frac{0.25(s + 40)}{s + 10}$$

Solution: Using the BLT method gives the digital filter $T(z)$, as shown next.

$$T(z) = \frac{0.2857z - 0.19046}{z - 0.90476}$$

In order to put the numerator in the standard form we used for poles, we pull out the a_0 from both terms in the numerator to get the following equation.

$$T(z) = \frac{0.2857(z - 0.6666)}{z - 0.90476}$$

The term in the zero factor is really a_1/a_0. It is this value that is used for c in Equation 9.2. A 4 percent reduction in this value of 0.6666 gives the new c as 0.6400. Using this value in Equation 9.2 for c gives the corner frequency of the zero of the new analog filter represented by the IIR difference equation with the new a coefficient, as shown next.

$$a = \frac{-\ln(0.6399)}{0.01} = 44.6$$

This is an 11.5 percent change in the zero corner frequency for a 4 percent change in the a coefficient ratio of the IIR difference equation.

9.3 Computing DSP Coefficient Word Length Effects

Equation 9.2 is useful for quantitatively computing the effect of changes in the b coefficients of the digital filters difference equation and the change produced in the analog filter corner frequency represented by the changed digital IIR filter. The equation even gives a good qualitative estimate of the maximum change allowed in any b coefficient of digital filters representing multiple pole analog filters. But the equation is hard to use, since a change in b is assumed and then the change in a, the corner frequency, is computed and checked to see if it is near the maximum percentage change allowed. What is needed is an equation where the specification on the corner frequency a is used to compute the maximum allowed change in the b coefficient. This is done in this section using first-semester calculus.

Let's start with Equation 9.2, where c is the pole or zero in the z-plane, by taking the derivative of both sides of the equation with respect to c. This results in the following equations, which result in Equation 9.4.

$$\frac{da}{dc} = -\frac{1}{T}(\frac{1}{c})$$

$$da = -\frac{dc}{Tc}$$

$$\left(\frac{da}{a}\right) = -\frac{1}{aT}\left(\frac{dc}{c}\right)$$

$$\left(\frac{dc}{c}\right)100 = -aT\left(\frac{da}{a}\right)100 \qquad \text{(Equation 9.4)}$$

In Equation 9.4, if the infinitesimal differentials were approximated by small increments in c and a, then for single-pole or zero analog filters the percentage change in the corner frequency of the pole or zero gives the percentage change in the corresponding difference equation coefficient. Remember that for the a coefficients, Equation 9.4 gives a_1/a_0. Equation 9.4 is valid for small percentage changes in c or a and for analog filters with single poles or zeros. For multiple poles or zeros, Equation 9.4 gives qualitative results that are valid within an order of magnitude. Example 9.8 illustrates the use of Equation 9.4.

Example 9.8. Using Equation 9.4 to find the b coefficient precision for a specified pole corner frequency accuracy

Problem: For the following single-pole analog LPF $T(s)$, determine the precision required to ensure that the difference equation of the corresponding IIR digital filter will have a corner frequency of the same accuracy for a sampling period $T = 0.002$ s. Let the analog filter require an accuracy of 5 percent for the corner frequency.

$$T(s) = \frac{50}{s+50}$$

Solution: The analog filter has a corner frequency of 50 rad/s, and a sampling period $T = 0.002$ s, while the percentage accuracy for the corner frequency is 5 percent. Using these values in Equation 9.4 gives the following equation.

$$\left(\frac{db}{b}\right)100 = -(0.1)(5) = -0.5 \text{ percent}$$

This says that the b_1 coefficient (the only b coefficient) must not vary from the computed value by more than 0.5 percent in order for it to represent an analog filter with a pole with a corner frequency of 50 rad/s within 5 percent. In order to check this, the BLT method is used to determine the IIR digital filters for the given $T(s)$ and one with a 5 percent change in the corner frequency. Then the difference equations for both will be obtained. It will be seen that the b coefficients of the two difference equations differ by 0.5 percent. This is done, with the first $T(z)$ corresponding to the preceding $T(s)$.

$$T(z) = \frac{4.761 * 10^{-2}(z+1)}{z - 0.90475}$$

$$T(z) = \frac{4.988 * 10^{-2}(z+1)}{z - 0.90023}$$

$$y(n) = 0.90475 y(n-1) + 0.04761 x(n) + 0.04761 x(n-1)$$

$$y(n) = 0.90023 y(n-1) + 4.988 * 10^{-2}[x(n) + x(n-1)]$$

The last difference equation has its b coefficient (0.90023) reduced by 0.5 percent from the difference equation above it. This is exactly what Equation 9.4 predicted.

Summary

In this chapter we looked at the precision required in terms of percentage change allowed in the b_1 coefficient of the difference equation of the IIR filter to ensure it represents the corresponding analog filter within its specification. The percentage change is due to the number of decimal digits or binary bits used to code the coefficient, as was discussed in Section 9.1. The equation that resulted was Equation 9.4, which gives the percentage change of the b coefficient in terms of the percentage change of the corner frequency of the analog pole allowed multiplied by the negative of the sampling period times the analog corner frequency. This result is applicable to the percentage change allowed for the ratio of a_1 over a_0 for the coefficient change for a specified analog zero corner frequency percentage change.

Equation 9.4 is valid only for single-zero or single-pole analog corner frequencies, or for single-pole and single-zero analog filters. However, it gives good qualitative results for multiple-pole or zero analog filters when

Digital Signal Processing

used for the pole or zero with the lowest corner frequency. This approximation improves when the sampling frequency is much higher than the corner frequencies. The major result is that many times the percentage error allowed in the coefficients of an IIR digital filter may be much less than that of the analog filter corner frequency specifications to ensure that the digital IIR filter performs to the specifications of the analog filter it replaces.

Self-Test

1. Determine the minimum number of bits to the right of the binary point to ensure that the decimal number 0.175 is coded with a precision of 5 percent or less.

2. Determine the minimum number of bits to the right of the binary point to ensure that the decimal number 2.105 is coded with a precision of 2 percent or less.

3. Determine the percentage change in the b_1 coefficient of the difference equation that would cause a 4 percent change in the corresponding lowpass analog filter $T(s)$ corner frequency it represents for a sampling time $T = 0.04$ s.

$$T(s) = \frac{25}{s + 25}$$

4. Determine the percentage change in the b_1 coefficient of the difference equation that would cause the highpass analog filter $T(s)$ it replaces to have a change in its corner frequency of 10 percent for the sampling period of $T = 0.005$ s.

$$T(s) = \frac{s}{s + 100}$$

5. Determine the percentage change in the b_1 coefficient of the difference equation that would cause the lowpass analog filter $T(s)$ it replaces to have a change in its pole corner frequency of 5 percent for $T = 0.002$ s.

$$T(s) = \frac{0.1(s + 100)}{s + 10}$$

Filter Coefficient Precision

6. Determine the percentage change in the ratio of a_1/a_0 of the difference equation that would cause the analog filter $T(s)$ represented by the following IIR digital filter to have a 1 percent change in its zero corner frequency for $T = 0.002$ s.

$$T(s) = \frac{0.1(s + 100)}{s + 10}$$

Problems

1. Determine the minimum number of bits to the right of the binary point to ensure that the decimal number 0.915 is coded with a precdision of 2 percent or less.

2. Determine the minimum number of bits to the right of the binary point to ensure that the decimal number 5.12 is coded with a precision of 1 percent or less.

3. Determine the percentage change in the b_1 coefficient of the difference equation that would cause a 5 percent change in the corresponding lowpass analog filter $T(s)$ corner frequency it represents for a sampling time $T = 0.02$ s.

$$T(s) = \frac{100}{s + 100}$$

4. Determine the percentage change in the b_1 coefficient of the difference equation that would cause the highpass analog filter $T(s)$ it replaces to have a change in its corner frequency of 2 percent for the sampling period $T = 0.003$ s.

$$T(s) = \frac{s}{s + 50}$$

5. Determine the percentage change in the b_1 coefficient of the difference equation that would cause the lowpass analog filter $T(s)$ it replaces to have a change in its pole corner frequency of 10 percent with $T = 0.001$ s.

$$T(s) = \frac{200}{s + 200}$$

6. Determine the percentage change in the ratio of a_1/a_0 of the difference equation that would cause the analog filter represented by the following IIR digital filter to have a 2 percent change in its zero corner frequency for the sampling period $T = 0.05$ s.

$$T(s) = \frac{s + 10}{s + 30}$$

Answers to Self-Test

1. 6

2. 4

3. 4 percent

4. 5 percent

5. 0.1 percent

6. 0.2 percent

chapter 10

FIR Filter Design

Introduction

In this chapter we learn how to design a new type of digital filter, the FIR filter. This filter has no analog filter equivalent. Thus the design methods of Chapter 7 are not applicable. In that chapter we started with a known analog filter, since the design methods for analog filters are well known, and used methods to convert the analog filter to a digital filter. These methods all resulted in Infinite Impulse Response (IIR) filters. The FIR filters are digital filters with no b coefficients with which to multiply previous outputs. Thus, no previous outputs need to be saved or used for computing. It seems reasonable that there must be a price to pay for this, and there is. Usually the FIR filter uses many more a coefficients and corresponding input samples than are saved by not using the b coefficients and previous outputs. Many times this is not much of a problem.

Because no b coefficients are used for FIR filters, there are no poles of the z transfer function describing the digital filter. And since we have seen that the z-plane is just a modified s-plane, using the relationship $z = e^{sT}$, we see that there would be no analog poles as there are in most analog filters. The advantages of the FIR filter is that the phase can be specified as linear, which is an advantage in some applications. In analog filters the phase corresponding to the analog filter that meets your specifications is just accepted; now it can be designed in, and usually it is chosen to be linear. The method of designing FIR filters is given in the following discussion along with some of the limitations of FIR filters. In Chapter 11, we introduce windows, which alleviate some of the effects of not using the b coefficients along with the previous outputs.

FIR filters are used in many applications where the advantages of linear phase are needed. Since IIR filters have analog filter equivalents, linear phase is more difficult to achieve. Their phase is determined by the placement and order of their poles and zeros. At the end of Chapter 11, Application 2 illustrates designing and analyzing an FIR filter to determine the RMS values of signals in a selected band of frequencies, as is done on most stereo equipment. This application could take advantage of the linear phase of the FIR filters to do graphical equalization of recorded sound instead of just displaying its RMS value. Some of the technical areas in which FIR filters are employed include speech recognition and enhancement, audio recording and equalization, telecommunications, signal and data smoothing, and ultrasound imaging.

10.1 Introduction to the FIR Filter

As was stated in the beginning of this chapter, we are dealing now with a digital filter that uses no b coefficients in its difference equation. This type of filter is called an **FIR filter**, for Finite Impulse Response filter. After a finite amount of time after an impulse signal is applied, the output of this type of digital filter goes to zero and stays there. The modified DSP equation for an FIR filter is obtained by leaving off the b coefficients and adding a coefficients with negative subscripts, as shown in Equation 10.1.

$$y(n) = a_0 x(n) + a_1 x(n-1) + \cdots + a_N x(n-N)$$
$$+ a_{-1} x(n+1) + \cdots + a_{-N} x(n+N) \quad \text{(Equation 10.1)}$$

This looks like a simple equation for a digital filter; all that needs to be done is derive the a coefficients to meet the filter specifications. However, a new method needs to be developed to do this, since the FIR filter has no poles and thus has no corresponding analog filter. These characteristics are derived as follows.

First, let's get the math description of the FIR filter, $T(z)$, by taking the z-transform of Equation 10.1, using the shifting property, and then solving for the output over the input, producing Equation 10.2.

$$Y(z) = a_0 X(z) + a_1 z^{-1} X(z) + \cdots + a_N z^{-N} X(z) + a_{-1} z X(z) + \cdots + a_{-N} z^N X(z)$$

$$T(z) = \frac{Y(z)}{X(z)} = a_0 + a_1 z^{-1} + \cdots + a_N z^{-N} + a_{-1} z + \cdots + z^N a_{-N} \quad \text{(Equation 10.2)}$$

FIR Filter Design

It can be seen in Equation 10.2 that there are no roots of the denominator, since there is no denominator. Also, we have seen in Chapters 8 and 9 that there is a correspondence between the s-plane poles and the z-plane poles, given by $z = e^{sT}$. Thus any corresponding analog filter given by $T(s)$ would have no s-plane poles, but this is limited to only a few analog filters. Yet we will see that any filter specification can be met by FIR filters; thus essentially there are no corresponding analog filters. The design methods must be derived from the FIR filter specifications, instead of using a corresponding analog filter.

10.2 The General FIR Coefficient Equation

We must derive the a coefficients from the digital filter frequency specifications, since there are no analog filters to convert to digital filters. We already have a method to determine the frequency response of a digital filter given the transfer function of the filter $T(z)$. This method is given in Equation 6.4, using $T(z)$. We have just found $T(z)$ for the FIR digital filter in Equation 10.2. If we knew the a coefficients, we could use Equation 10.2 to get the $T(z)$ for a specific filter corresponding to those coefficients. Then we could use Equation 6.4 to find its frequency response or gain plot. What we now need to do is reverse the preceding steps and determine the coefficients from the frequency specifications; that is, solve Equation 6.4 for the coefficients.

Let's begin by rewriting Equation 10.2 in a more convenient mathematical form, using the summation notation to get the following equation.

$$T(z) = \sum_{k=-N}^{N} a_k z^{-k}$$

Let's also replace z by e^{jwT} in Equation 6.3 to get the following equation for the frequency response of the FIR filter, where as before $\Omega = wT$.

$$T(e^{j\Omega}) = \sum_{k=-N}^{N} a_k e^{-jk\Omega}$$

If the magnitude was taken of both sides, the preceding equation would give the gain, as was done in Equation 6.4 for any digital filter.

Now the left side of the preceding equation is the specified frequency response, which is known. If it can be solved for the *a* coefficients, they can be used in Equation 10.1 to write the difference equation of the desired FIR filter. The preceding equation can be solved for the *a* coefficients, as shown in the following equations.

$$T(e^{j\Omega}) = \sum_{k=-N}^{N} a_k e^{-j\Omega k}$$

$$T(e^{j\Omega})e^{j n\Omega} = \sum_{k=-N}^{N} a_k e^{j\Omega(n-k)}$$

$$\int_0^{2\pi} T(e^{j\Omega})e^{j n\Omega} d\Omega = \sum_{k=-N}^{N} \int_0^{2\pi} a_k e^{j\Omega(n-k)} d\Omega$$

$$= \sum_{k=-N}^{N} \int_0^{2\pi} a_k [\cos(n-k)\Omega + j\sin(n-k)\Omega] d\Omega$$

$$= \sum_{k=-N}^{N} [\int_0^{2\pi} a_k \cos(n-k)\Omega d\Omega + \int_0^{2\pi} j a_k \sin(n-k)\Omega d\Omega]$$

Now Ω goes over 2π, and the integral of any sine or cosine over $(n-k)2\pi$ is always zero, except for $n-k=0$. The preceding equation becomes the following equation, where $k = n$.

$$\int_0^{2\pi} T(e^{j\Omega})e^{j\Omega n} d\Omega = \int_0^{2\pi} a_n \cos(0)\Omega d\Omega + \int_0^{2\pi} j a_n \sin(0)\Omega d\Omega$$

$$= \int_0^{2\pi} a_n d\Omega = 2\pi a_n$$

Thus we get Equation 10.3, which computes the *a* coefficients, given the frequency response specification $T(e_j^{\Omega})$ of the FIR filter, with $\Omega = wT$.

FIR Filter Design

$$a_n = \frac{1}{2\pi} \int_0^{2\pi} T(e^{j\Omega}) e^{jn\Omega} d\Omega \qquad \text{(Equation 10.3)}$$

Example 10.1 shows how to use Equation 10.3 to find the FIR a coefficients, given the filter frequency specification.

Example 10.1. Finding the a_0 and a_2 FIR filter coefficients for an LPF

Problem: Let the frequency specification of the LPF be the gain = 1 to 25 Hz, and gain = 0 beyond. Also let the sampling period $T = 0.002$ s.

Solution: Since $\Omega = wT$, we have that the gain is 1 from $\Omega = 0$ to 0.1π and from 1.9π to 2π, since the frequency spectrum is repeated and symmetrical about π. Using this in Equation 10.3 gives the following equations.

$$a_n = \frac{1}{2\pi} \int_0^{0.1\pi} e^{jn\Omega} d\Omega + \frac{1}{2\pi} \int_{1.9\pi}^{2\pi} e^{jn\Omega} d\Omega$$

$$= \frac{1}{2\pi} \int_{-0.1\pi}^{0.1\pi} e^{jn\Omega} d\Omega$$

Evaluating the integral gives the following equations for n not equal to zero, and $n = 0$.

$$a_n = [\frac{e^{jn\Omega}}{jn2\pi}]_{-0.1\pi}^{0.1\pi}, \text{ for } n \text{ not equal zero}$$

$$a_0 = 0.1$$

Using $n = 2$ in the first equation gives the following equations for a_2.

$$a_2 = \frac{e^{0.2\pi j} - e^{-0.2\pi j}}{2\pi(2j)}$$

$$a_2 = \frac{\cos(0.2\pi) + j\sin(0.2\pi) - \cos(-0.2\pi) - j\sin(-0.2\pi)}{2\pi(2j)}$$

$$a_2 = \frac{\sin(0.2\pi)}{2\pi}$$

Digital Signal Processing

The final equation for a_2 was derived using the fact that $\sin(-a) = -\sin(a)$, and $\cos(-a) = \cos(a)$.

In Example 10.1, the fact that all digital filters and filter specifications are periodic, with a period of the sampling frequency, allowed the integration to be simplified by changing the limits to give a similar but simpler integral to be evaluated. This can be done for other filters, too. All that is required is that Equation 10.3 be integrated over 2π radians. Since solving Equation 10.3 is tedious, we solve it for the four basic ideal types of filters with arbitrary pass and stop frequencies in the next section.

10.3 The Basic Solutions of the Coefficient Equation

The ideal lowpass digital filter specification is given in Figure 10.1. Using the values given in Figure 10.1 in Equation 10.3 gives the following integral equation and its algebraic solution, where Ω_p is the ideal pass and stop frequency with no transition region between them.

$$a_n = \frac{1}{2\pi} \int_{-\Omega_p}^{\Omega_p} e^{jn\Omega} d\Omega$$

$$a_0 = \frac{\Omega_p}{\pi} \qquad \text{(Equation 10.4a)}$$

$$a_n = \frac{\sin(n\Omega_p)}{n\pi}, \text{ for } n \text{ not equal to } 0 \qquad \text{(Equation 10.4b)}$$

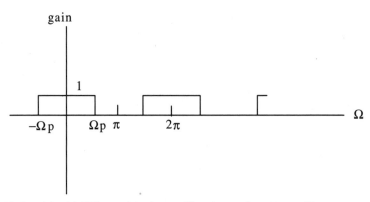

Figure 10.1. Ideal LPF graphical specification using $\Omega = wT$

FIR Filter Design

The ideal highpass digital filter specification is given in Figure 10.2. Using the values and symbols in Figure 10.2 in Equation 10.3 gives the following integral equation and its algebraic solutions, where again Ω_p is the ideal pass and stop frequency with no transition region.

$$a_n = \frac{1}{2\pi} \int_{\Omega_p}^{2\pi - \Omega_p} e^{jn\Omega} d\Omega$$

$$a_0 = \frac{\pi - \Omega_p}{\pi} \qquad \text{(Equation 10.5a)}$$

$$a_n = \frac{-\sin(n\Omega_p)}{\pi n}, \text{ for } n \text{ not equal to zero} \qquad \text{(Equation 10.5b)}$$

The ideal bandpass filter is shown in Figure 10.3. Using the values shown in Figure 10.3 in Equation 10.3 gives the following integral equation and algebraic solutions, where $\Omega 1$ is the lower passband and stopband frequency, and $\Omega 2$ is the upper passband and stopband frequency. Again, both transition bands do not exist.

$$a_n = \frac{1}{2\pi} \int_{\Omega 1}^{\Omega 2} e^{jn\Omega} d\Omega + \frac{1}{2\pi} \int_{-\Omega 2}^{-\Omega 1} e^{jn\Omega} d\Omega$$

$$a_0 = \frac{\Omega 2 - \Omega 1}{\pi} \qquad \text{(Equation 10.6a)}$$

$$a_n = \frac{1}{\pi n}[\sin(n\Omega 2) - \sin(n\Omega 1)], \text{ for } n \text{ not equal to zero.} \qquad \text{(Equation 10.6b)}$$

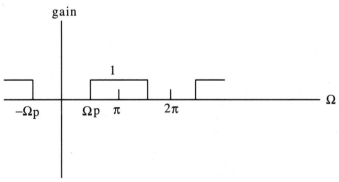

Figure 10.2. Ideal HPF graphical specification using $\Omega = wT$

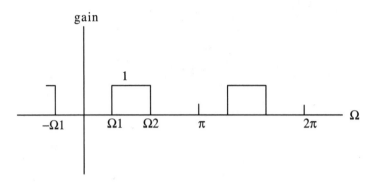

Figure 10.3. Ideal BPF graphical specification using $\Omega = wT$

The ideal bandstop filter specification is shown in Figure 10.4. Using the values and variables shown in Figure 10.4 in Equation 10.3 gives the following integral equation and algebraic solutions, where again the lower passband and stopband frequency is $\Omega 1$ and the upper passband and stopband frequency is $\Omega 2$, with no transition regions.

$$a_n = \frac{1}{2\pi} \int_{-\Omega 1}^{\Omega 1} e^{jn\Omega} d\Omega + \frac{1}{2\pi} \int_{\Omega 2}^{2\pi - \Omega 2} e^{jn\Omega} d\Omega$$

$$a_0 = \frac{\pi + \Omega 1 - \Omega 2}{\pi} \qquad \text{(Equation 10.7a)}$$

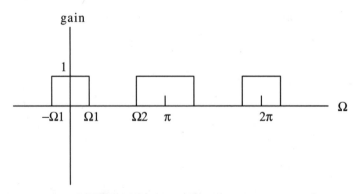

Figure 10.4. Ideal BSF graphical specification using $\Omega = wT$

FIR Filter Design

$$a_n = \frac{1}{n\pi}[\sin(n\Omega1) - \sin(n\Omega2)], \text{ for } n \text{ not equal to zero} \quad \text{(Equation 10.7b)}$$

These various algebraic equations can be used to solve for the a coefficients of the four basic ideal digital filter specifications. Example 10.2 does this for the first two coefficients of the lowpass filter in Example 10.1, and Example 10.3 does this for a bandpass filter.

Example 10.2. Finding the first two FIR coefficients with nonnegative subscripts for the lowpass filter of Example 10.1

Problem: The lowpass filter gain is given as 1 from 0 to 25 Hz, and zero beyond 25 Hz. The sampling period is 0.002 s.

Solution: Using the sampling period of 0.002 s gives $\Omega_p = 0.002(2\pi)25 = 0.1\pi$, which gives the following equations for the first two coefficients.

$$a_0 = 0.1$$

$$a_1 = \frac{\sin(0.1\pi)}{\pi}$$

Example 10.3. Finding the first two coefficients with nonnegative subscripts for an ideal FIR bandpass filter

Problem: Let the gain of the filter be 1 between 25 and 100 rad/s and zero everywhere else. The sampling period is $T = 0.01$ s.

Solution: For a sampling period of 0.01 s, the values of the passband frequencies are computed as follows.

$$\Omega1 = 0.01(25) = 0.25$$

$$\Omega2 = 0.01(100) = 1.00$$

Using these values in the algebraic solutions for the bandpass FIR coefficients gives the following values.

$$a_0 = \frac{1.00 - 0.25}{\pi} = \frac{0.75}{\pi}$$

$$a_1 = \frac{1}{\pi}[\sin(1) - \sin(0.25)] = \frac{0.594}{\pi}$$

Examples 10.2 and 10.3 show how easy it is to compute the a coefficients for a digital FIR filter, using the algebraic solutions of Equation 10.3 for the four basic ideal filters. However, the number of coefficients to compute was left undetermined, except that Equation 10.3 was developed for as many coefficients with negative subscripts as positive. All that can be said at this point is that the filter coded using a difference equation with more coefficients is closer to the ideal filter than one that is not. This means that the transitions regions get narrower, and the ripple in the passband and stopbands is reduced as the number of coefficients used is increased.

10.4 Use of the Basic Solutions

Since there is no analog filter that corresponds to the FIR digital filter, FIR filter design using Equations 10.4 through 10.7 requires the designer to start from the ideal filter graphical specifications. This procedure is shown using Mathcad in Example 10.4 for a lowpass filter. The use of Mathcad is not necessary; it is used only to speed up repetitive calculations, which could be done on a calculator or another mathematical program.

From the graphical specifications the corner frequencies of interest are determined. Since the ideal filters have no transition regions, the highest frequency in the passband corresponds to the lowest frequency in the stopband for a lowpass filter. This is also the case for the upper stopband for a passband filter or the lower passband for a stopband filter. Similarly, the lowest frequency in the passband corresponds to the highest frequency in the stopband for a highpass filter. This is also the case for the lower stopband for a passband filter or the upper passband for a stopband filter. These corner frequencies are then multiplied by the sampling period T and used in the appropriate equation, using Equations 10.4 through 10.7, where $\Omega = wT$.

Notice that in Equations 10.4 through 10.7 the coefficients are the same for negative or positive subscripts. These are called *symmetrical coefficients*, and thus only about half the coefficients need to be computed. This form of coefficients gives the most widely applicable FIR filters. Next,

FIR Filter Design

the coefficients are multiplied by their corresponding sampled input signal as shown in Equation 10.1 and summed to get the current output signal $y(n)$. Notice that in Equation 10.1 the input samples with arguments greater than n correspond to samples that have not occurred yet. This gives the *noncausal form* of the FIR filter and is not a problem if the input samples consist of previously stored data. However, for real-time FIR filters the output value corresponding to $y(n)$ must be delayed until all the input samples needed are available. This leads to the *causal form* of the FIR filter, which is discussed in section 10.5. For now, we will proceed with the design of the noncausal FIR ideal filter.

The only thing not known in the preceding discussion is the number of positive subscripted coefficients to compute. Using mathematical programs to compute the coefficients and plot the frequency response using the methods in Chapter 6 alleviates this problem. In Example 10.4 using Mathcad, the number of coefficients with positive subscripts N and thus the filter length $2N + 1$ are easily determined by seeing if the plotted gain curve meets the original graphical specification. If it does not, the value of N is increased. The design of an ideal lowpass FIR filter is shown in Example 10.4 using Mathcad. Figure 10.5 is the gain plot in dB for $N = 4$ used in Example 10.4. If the value of N in Example 10.4 is increased to 8, then the gain plot in dB in Figure 10.6 is obtained. The filter plotted in Figure 10.6 has a narrower transition region and lower ripple in the passband and stopband.

Example 10.4. Using Mathcad to Design and Check FIR LPF Design

$N := 4$

$n := 1 .. N$ The positive coefficient subscripts

$T := 0.01$ The sampling period in s

$w_p := 100$ rad/s, The lowpass frequency

$a_0 := w_p \cdot \dfrac{T}{\pi}$ The zero coefficient

$a_n := \dfrac{\sin(n \cdot w_p \cdot T)}{n \cdot \pi}$ The FIR coefficients

$k := 0..30$ The number of data points to use to plot the gain

$w_k := 10 \cdot k$ The data point number converted to frequency in rad/s

$$z_k := \cos(w_k \cdot T) + \sin(w_k \cdot T) \cdot j$$

$$Tz_k := ab_0 + \sum_n ab_n \cdot \left[(z_k)^n + (z_k)^{-n} \right]$$

$$gain_k := |Tz_k|$$

$$gaindB_k := 20 \cdot \log(gain_k)$$

10.5 The Causal (Real Time) and Noncausal Filter Coefficients

We have found a way to design an FIR digital filter of the form of Equation 10.1. The problem is that this is the input-output difference equation for a **noncausal** digital filter. *This means that it doesn't follow the usual cause then effect process*, in that the output occurs before all the input needed is available. This can be seen by looking at the inputs with arguments greater than n. These arguments mean that they are input samples taken later than the nth sample, but the nth input and output samples are the current samples. As an example, $x(n+1)$ is the sample value of the input taken one sample later than the time corresponding to the output sample $y(n)$ that uses it. There is no problem in using noncausal digital filters if the input samples have been stored ahead of time, and then digital filtering is done on the stored data. This is illustrated in Example 10.5.

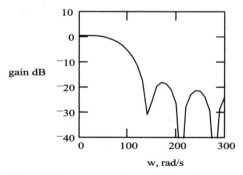

Figure 10.5. FIR lowpass filter gain for $N = 4$

FIR Filter Design

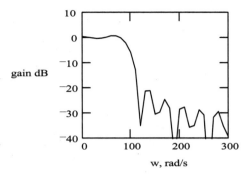

Figure 10.6. FIR lowpass filter gain for $N = 8$

Example 10.5. Using a noncausal filter on stored data

Problem: Use a very simple lowpass filter with $N = 1$ to smooth the following noisy data points stored in computer memory.

$x(0) = 1$

$x(1) = 0.7$

$x(2) = 1.1$

$x(3) = 1.3$

$x(4) = 0.8$

$x(5) = 1.1$

$x(6) = 0.9$

Solution: A very simple lowpass (smoothing) digital filter is given by the following equation, which is noncausal because $x(n + 1)$ is needed to compute the smoothed output value at the nth sample. This is no problem using the given stored data.

$$y(n) = 0.333x(n+1) + 0.333x(n) + 0.333x(n-1)$$

Using this FIR filter on the given data, we get the following smoothed output data.

$y(1) = 0.933$

$y(2) = 1.033$

$y(3) = 1.066$

$y(4) = 1.066$

$y(5) = 0.933$

It can be seen that there was no problem in computing the digital FIR filter outputs, except for the end points, where there was insufficient data. The smoothed signal out of the filter varied between 0.933 and 1.066, while the unfiltered input data varied between 0.7 and 1.3. If the variation on the input data was due to noise, then the smoothed output data is a better representation of a slowly varying input signal that is now around 1.

However, if the input signal is just coming in and the current value of the digital FIR filter is needed at the time corresponding to its sample number, there is a problem. The only solution is to delay the computation of the output until all the data is available. Thus the output value is delayed by N samples and the value is correct, but it is nT seconds late. In Example 10.5, if the input signal was just coming into an A/D, the output sample would be delayed by T seconds until the next sample after the current one was received to use in the computation. There is no way around this effect, but to make the computation easier the digital FIR filter is written in the causal form by shifting the sample numbers of the input signal and their corresponding subscripts so that they start at the current sample number n and count down to previous sample numbers. This reduction of the input sample arguments by N and changing the coefficient subscripts to correspond to the new arguments give the **causal form** of the FIR filter. This is not a solution, it just computes the output samples late by N sample periods. This is illustrated by redoing Example 10.5 in Example 10.6, using the causal form of the FIR filter.

Example 10.6. Using a causal lowpass filter to smooth noisy data

Problem: The noisy data is again given in Example 10.5, but now the original noncausal filter is changed into its causal form so that it can work on the input signal as it comes in.

FIR Filter Design

Solution: The original noncausal filter is repeated here.

$$y(n) = 0.333x(n+1) + 0.333x(n) + 0.333x(n-1)$$

Since $N = 1$ in the noncausal filter, the new causal equation is gotten by subtracting 1 from the arguments of the sampled signal. This gives the following causal equation.

$$y(n) = 0.333x(n) + 0.333x(n-1) + 0.333x(n-2)$$

This causal equation gives the following outputs when there is sufficient input data to compute the output.

$y(2) = 0.933$

$y(3) = 1.033$

$y(4) = 1.066$

$y(5) = 1.066$

$y(6) = 0.933$

When these values are compared to the noncausal filter outputs of Example 10.5, it is seen that they are the same except delayed by one sampling period. There is no way around requiring input samples that have not occurred in real-time filters except to wait until the data is available.

In Example 10.6 the coefficients were already converted to numerical values. If the coefficients are still symbolic, then the coefficient subscripts must be modified to correspond to the new arguments of the corresponding filter. Remember, all that is really done is to wait until the data needed is available so that the relationship of the coefficient to its sample is unchanged. This is illustrated in Example 10.7.

Example 10.7. Changing a general noncausal FIR filter equation to the causal form

Problem: Find the causal form of the following general noncausal FIR digital filter equation.

$$y(n) = a_{-2}x(n+2) + a_{-1}x(x+1) + a_0x(n) + a_1x(n-1) + a_2x(n-2)$$

Solution: In this noncausal equation, $N = 2$. If the arguments of the input sample are reduced by this, then all the data required is available, it's just that the output is late by two sample periods. The new causal form of the noncausal equation of the FIR digital filter is given by the following equation, after the coefficient subscripts are also changed.

$$y(n) = a_0 x(n) + a_1 x(n-1) + a_2 x(n-2) + a_3 x(n-3) + a_4 x(n-4)$$

From this equation it can be seen that what is being computed is really the value of $y(n-2)$, but it is being used for the $y(n)$ value. You can't change reality by renaming some values.

Summary

In this chapter we learned to design a new type of digital filter, the FIR filter. The input-output difference equation for this filter is given in Equation 10.1. It has no b coefficients, but has a coefficients with positive and negative subscripts. In this form the FIR filter is called noncausal because inputs that have not occurred are required to compute the output.

The a coefficients are solved for by equating the frequency response of the FIR filter transfer function $T(z)$ to the desired frequency response to get Equation 10.3. Equation 10.3 is solved for the four basic ideal digital filter graphical specifications in Figures 10.1 through 10.4 to get Equations 10.4 through 10.7 for the corresponding a coefficients. The method of transforming the noncausal FIR difference equation into a causal form with a delay of NT seconds is given in Section 10.5. The arguments of the input samples are reduced by N, and the coefficient subscripts are increased by N, so that no future samples are used.

Self-Test

1. Use Equation 10.4 to determine the a_0 and a_1 coefficients for an FIR lowpass digital filter with $\Omega_{p\,=\,0.25}\pi$.

2. Use Equation 10.4 to determine the a_2 and a_{-1} coefficients for an FIR lowpass digital filter with $\Omega_{p\,=\,0.5}\pi$.

FIR Filter Design

3. Use Equation 10.5 to determine the a_0 and a_3 coefficients for a highpass FIR digital filter with $\Omega_p = 1.25$.

4. Use Equation 10.5 to determine the a_{10} coefficient for a highpass FIR digital filter with a gain of zero out to 25 rad/s and then 1 thereafter. Let the sampling period $T = 0.02$ s.

5. Use Equation 10.6 to determine the a_1 coefficient for a bandpass FIR digital filter with $\Omega 1 = 0.5$ and $\Omega 2 = 1.0$.

6. Convert the following noncausal FIR filter to its causal form.
$$y(n) = 0.24x(n+2) + 0.5x(n+1) + x(n) + 0.5x(n-1) + 0.24x(n-2)$$

7. Convert the following noncausal FIR digital filter to its causal form.
$$y(n) = a_{-2}x(n+2) + a_{-1}x(n+1) + a_0 x(n) + a_1 x(n-1) + a_2 x(n-2)$$

8. Use Equation 10.7 to determine the a_0 and a_2 coefficients for a stopband FIR filter with $\Omega 1 = 0.2$ and $\Omega 2 = 1.2$.

9. Use Equation 10.6 to determine the a_0 and a_2 coefficients for a bandpass FIR filter with $\Omega 1 = 0.2$ and $\Omega 2 = 1.2$.

10. Determine the a_0 and a_1 coefficients for a lowpass FIR filter using Equation 10.4, where the desired ideal passband and stopband frequency is 25 rad/s and the sampling period $T = 0.01$ s.

11. Determine the a_0 and a_1 coefficients for a lowpass FIR filter using Equation 10.4, where the desired ideal passband and stopband frequency is 50 rad/s and the sampling period $T = 0.01$ s.

12. Use Mathcad with Equation 10.4 to determine the nine a coefficients from a_{-4} to a_4 for an FIR lowpass digital filter with $\Omega_p = 0.7\pi$.

13. Use Mathcad with Equation 10.6 to determine the eleven a coefficients from a_{-5} to a_5 for an FIR bandpass filter with $\Omega 1 = 1.0$ and $\Omega 2 = 1.5$.

14. Use Mathcad with Equation 10.5 to determine the nine a coefficients from a_{-4} to a_4 for an FIR highpass filter where the ideal passband and stopband frequency is 100 rad/s and the sampling period is $T = 0.017$ s.

Problems

1. Use Equation 10.4 to determine the a_{-2} and a_0 coefficients for an FIR lowpass filter with $\Omega_p = 0.5\pi$.

2. Use Equation 10.4 to determine the a_{-10} and a_5 coefficients for an FIR lowpass filter with $\Omega_p = 0.2\pi$.

3. Use Equation 10.5 to determine the a_0 and a_2 coefficients for a highpass FIR filter with $\Omega_p = 0.25\pi$.

4. Use Equation 10.5 to determine the a_{-4} coefficient for a highpass FIR filter with a gain of zero out to 50 rad/s and 1 thereafter for a sampling period $T = 0.03$ s.

5. Use Equation 10.6 to determine the a_4 coefficient for a bandpass FIR filter with $\Omega 1 = 2.1$ and $\Omega 2 = 2.9$.

6. Convert the following noncausal FIR filter to its causal form.

$$y(n) = 0.105x(n+3) + 0.205x(n+2) + 0.30x(n+1)$$
$$+ 0.35x(n) + 0.30x(n-1) + 0.205x(n-1) + 0.105x(n-3)$$

7. Convert the following noncausal FIR filter to its causal form.

$$y(n) = a_{-3}x(n+3) + a_{-2}x(n+2) + a_{-1}x(n+1)$$
$$+ a_0x(n) + a_1x(n-1) + a_2x(n-2) + a_3x(n-3)$$

8. Use Equation 10.7 to determine the a_0 and the a_5 coefficients for a bandstop FIR filter with $\Omega 1 = 1.2$ and $\Omega 2 = 2.2$.

9. Use Equation 10.6 to determine the a_0 and the a_{-5} coefficients for a bandpass FIR filter with $\Omega 1 = 1.2$ and $\Omega 2 = 2.2$.

10. Determine the a_0 and the a_2 coefficients for a lowpass FIR filter using Equation 10.4, where the desired ideal passband and stopband frequency is 100 rad/s and the sampling period $T = 0.02$ s.

FIR Filter Design

11. Determine the a_1 and a_4 coefficients for a highpass FIR filter using Equation 10.5, where the desired ideal passband and stopband frequency is 100 rad/s and the sampling period $T = 0.02$ s.

Answers to Self-Test

1. 0.25, 0.225

2. 0, 0.318

3. 0.602, 0.061

4. 0.031

5. 0.115

6. $y(n) = 0.24x(n) + 0.5x(n-1) + x(n-2) + 0.5x(n-3) + 0.24x(n-4)$

7. $y(n) = a_0x(n) + a_1x(n-1) + a_2x(n-2) + a_3x(n-3) + a_4x(n-4)$

8. 0.682, −0.0455

9. 0.314, 0.0455

10. 0.0796, 0.0788

11. 0.159, 0.152

12. $a_{-4} = 0.0468$, $a_0 = 0.7$

13. $a_{-5} = 0.121$, $a_0 = 0.159$

14. $a_{-4} = -0.393$, $a_0 = 0.459$

chapter 11

Windows for FIR Filters

Introduction

In this chapter we see that FIR filters have poor gain characteristics for ideal filters, which have a jump between one and zero. We have seen in Chapter 10 that as the number of coefficients was increased, the transition region for the actual FIR filter decreased and the ripple in the passbands and stopbands was reduced. However, there is a limit to the reduction in the amplitude of the ripple as the number of coefficients used is increased. This is called the Gibbs effect. This effect is due to the jump between zero and one of the ideal gain curves. In order to reduce this effect, the actual values of the FIR coefficients are reduced near where they start and end. This is called windowing. There is no one best window, and several are introduced in this chapter. Two other methods of reducing the Gibbs effect are also discussed briefly.

11.1 The Gibbs Effect

In Chapter 10 we learned to compute the a coefficients for the difference equation of an FIR digital filter. We saw in Figures 10.5 and 10.6 that as the number of coefficients computed and used increased, the gain of the filter approached the ideal lowpass filter. The transition region got smaller and the ripple in the passband and stop band was reduced. However the ripple was moved closer to where the jump in gain occurred in the ideal filter gain plot. In the limit, when very many coefficients are used, the ripple is very narrow and centered about the ideal filter jump frequency, and the ripple amplitude approaches 9 percent of the jump value. This is called the **Gibbs effect**. It can be seen in Figure 11.1, where the example used in Figure 10.5 and 10.6 is used again, but the number of positive subscripted coefficients N is now 20. This is just Example 10.4, a lowpass FIR filter, with forty-one coefficients and more sample points on the frequency axis. Also, gain instead of gain in dB is plotted to show the Gibbs effect in the passband more effectively. As can be seen by comparing Figures 10.5, 10.6, and 11.1, the gain in the lowpass filter is approaching the ideal shown in Figure 10.1, but the ripple is converging about $\Omega_p = 1$ and is approaching a limit of about 1.09 in amplitude.

In order to reduce the Gibbs effect, the coefficient magnitudes are reduced as their subscripts approach N and $-N$. This is called **windowing**, since the effect may be thought of as looking at the FIR filter coefficients through a window that reduces their amplitude at the edges of the window. After windowing, the new FIR filter coefficients are multiplied by the window coefficients. This is illustrated in Figure 11.2. In order to talk about windowing in general, FIR filter coefficients that have not been windowed are said to be rectangular windowed (with all window coefficients having a value of 1 from N to $-N$).

Windowing will cause the magnitude of the ripple to decrease but increase the width of the transition band. There are many ways to reduce the end coefficients, and several are given in this chapter. Some are easy to compute, while others are more complicated to compute. Engineering judgment is used to decide the best method to use for the particular FIR filter and its specifications. Because of the symmetry of the noncausal

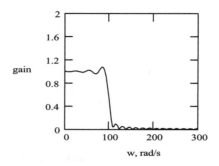

Figure 11.1. FIR lowpass filter gain for $N = 20$

Windows for FIR Filters

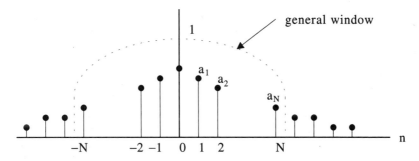

Figure 11.2. General windowing of FIR coefficients

coefficients, the window weighting factors are computed for the noncausal coefficients. Section 11.2 discusses the formulas for several windows and their advantages and disadvantages.

11.2 Several Windows

The Bartlett Window

The easiest window coefficients to compute are the Bartlett coefficients. They simply are weighting factors for the noncausal FIR filter coefficients that linearly decrease from 1 for multiplying a_0 to 0 for multiplying the a_N coefficient. Since the Bartlett and noncausal FIR filter coefficients are symmetrical, only the Bartlett coefficients for the a coefficients subscripted from 0 to N need to be computed. Equation 11.1 gives the formula for computing the Bartlett coefficients, and a plot of these values is given in Figure 11.3.

$$k_n = k_{-n} = \frac{N-n}{N} \text{ for } n \text{ between 0 and } N \quad \text{(Equation 11.1)}$$

As can be seen in Figure 11.3, the plot of the Bartlett coefficients is a triangle going from $-N$ to N with the values going from 0 to 1 and back to 0 again. The Bartlett window coefficients are the easiest to compute, but they give the widest transition band for the given amount of ripple reduction. Example 11.1 shows the computation and use of the Bartlett window coefficients for Example 10.4, with $N = 20$ as shown in Figure 11.1 without windowing. The new windowed filter gain is plotted in Figure 11.4.

Digital Signal Processing

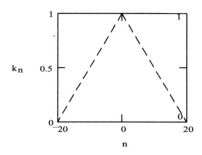

Figure 11.3. Bartlett window coefficients for $N = 20$

Example 11.1. Computing and using the Bartlett window coefficients on a lowpass FIR filter

Problem: Given the lowpass filter specifications in Example 10.4, compute the Bartlett window coefficients using Equation 11.1 for $N = 20$. Then apply them to the noncausal filter coefficients to produce the windowed noncausal filter coefficients, and plot the gain using Mathcad.

Solution:

$N := 20$

$n := 1 .. N$ The positive coefficient subscripts

$T := 0.01$ The sampling period in s

$w_p := 100$ rad/s   The lowpass frequency

$a_0 := w_p \cdot \dfrac{T}{\pi}$ The zero coefficient

$a_n := \dfrac{\sin(n \cdot w_p \cdot T)}{n \cdot \pi}$ The FIR coefficients

$k := 0 .. 100$ The number of data points to use to plot the gain

$w_k := 3 \cdot k$ The data point number converted to frequency in rad/s

$z_k := \cos(w_k \cdot T) + \sin(w_k \cdot T) \cdot j$

Windows for FIR Filters

$bk_0 := 1$

$bk_n := \dfrac{N - n}{N}$ The Bartlett coefficients for positive n

$ab_0 := a_0 \cdot bk_0$ The noncausal Bartlett windowed coefficient for n = 0

$ab_n := a_n \cdot bk_n$ The noncausal Bartlett windowed coefficients for positive n

$$Tz_k := ab_0 + \sum_n ab_n \cdot \left[(z_k)^n + (z_k)^{-n} \right]$$

$gain_k := |Tz_k|$

n	a_n	bk_n	ab_n
1	0.268	0.95	0.254
2	0.145	0.9	0.13
3	0.015	0.85	0.013
4	-0.06	0.8	-0.048
5	-0.061	0.75	-0.046
6	-0.015	0.7	-0.01
7	0.03	0.65	0.019
8	0.039	0.6	0.024
9	0.015	0.55	$8.017 \cdot 10^{-3}$
10	-0.017	0.5	$-8.658 \cdot 10^{-3}$
11	-0.029	0.45	-0.013
12	-0.014	0.4	$-5.693 \cdot 10^{-3}$
13	0.01	0.35	$3.601 \cdot 10^{-3}$
14	0.023	0.3	$6.757 \cdot 10^{-3}$
15	0.014	0.25	$3.45 \cdot 10^{-3}$
16	$-5.728 \cdot 10^{-3}$	0.2	$-1.146 \cdot 10^{-3}$
17	-0.018	0.15	$-2.7 \cdot 10^{-3}$
18	-0.013	0.1	$-1.328 \cdot 10^{-3}$
19	$2.511 \cdot 10^{-3}$	0.05	$1.255 \cdot 10^{-4}$
20	0.015	0	0

From Figure 11.4 it can be seen that using the Bartlett window on the lowpass FIR filter for $N = 20$ has reduced the ripple but increased the transition region compared to the same unwindowed lowpass FIR filter, which was plotted in Figure 11.1. In many cases this reduction in ripple is preferable to the increase in the transition region.

Digital Signal Processing

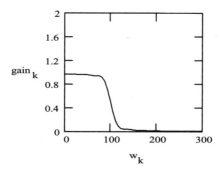

Figure 11.4. Bartlett windowed FIR lowpass filter gain for N = 20

The von Hann Window

Another window is the von Hann, which requires more complex computation of the coefficients, but causes less widening of the transition region. The coefficients for the von Hann window are computed using Equation 11.2.

$$k_n = k_{-n} = 0.5\{1 - \cos[\frac{\pi(N-n)}{N}]\}$$ (Equation 11.2)

By looking at Equation 11.2 it can be seen that the von Hann coefficients go from 0 to 1 and back to 0 as n goes from $-N$ to 0 and then to N. This is the same thing that the Bartlett coefficients do, but the von Hann coefficients smooth out the changes in slope of the lines going through the window coefficients so that there is no break at $-N$, 0, and N, as there was with the Bartlett coefficients. This is shown in Figure 11.5. Example 11.2 uses the von Hann window on the same lowpass filter used in Example 11.1. The resulting gain is plotted in Figure 11.6.

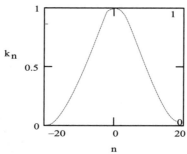

Figure 11.5. Von Hann window coefficients for N = 20

Windows for FIR Filters

Example 11.2. Computing and using the von Hann window coefficients on a lowpass FIR filter

Problem: Use Equation 11.2 to compute the von Hann window coefficients for the lowpass FIR filter used in Example 11.1 with $N = 20$ again. This is the same ideal lowpass filter used in Example 10.4 with $w_p = 100$ and the sampling period $T = 0.01$.

Solution:

$N := 20$

$n := 1 .. N$ The positive coefficient subscripts

$T := 0.01$ The sampling period in s

$w_p := 100$ rad/s   The lowpass frequency

$a_0 := w_p \cdot \dfrac{T}{\pi}$ The zero coefficient

$a_n := \dfrac{\sin(n \cdot w_p \cdot T)}{n \cdot \pi}$ The FIR coefficients

$k := 0 .. 100$ The number of data points to use to plot the gain

$w_k := 3 \cdot k$ The data point number converted to frequency in rad/s

$z_k := \cos(w_k \cdot T) + \sin(w_k \cdot T) \cdot j$

$vk_0 := 1$

$vk_n := 0.5 \cdot \left[1 - \cos\left[\dfrac{\pi \cdot (N - n)}{N}\right]\right]$ The von Hann coefficients for positive n

$av_0 := a_0 \cdot vk_0$ The noncausal von Hann windowed coefficient for n = 0

$av_n := a_n \cdot vk_n$ The noncausal von Hann windowed coefficients for positive n

$Tz_k := av_0 + \sum_n av_n \cdot \left[(z_k)^n + (z_k)^{-n}\right]$

$gain_k := |Tz_k|$

Digital Signal Processing

n	a_n	vk_n	av_n
1	0.268	0.994	0.266
2	0.145	0.976	0.141
3	0.015	0.946	0.014
4	-0.06	0.905	-0.054
5	-0.061	0.854	-0.052
6	-0.015	0.794	-0.012
7	0.03	0.727	0.022
8	0.039	0.655	0.026
9	0.015	0.578	$8.428 \cdot 10^{-3}$
10	-0.017	0.5	$-8.658 \cdot 10^{-3}$
11	-0.029	0.422	-0.012
12	-0.014	0.345	$-4.917 \cdot 10^{-3}$
13	0.01	0.273	$2.809 \cdot 10^{-3}$
14	0.023	0.206	$4.642 \cdot 10^{-3}$
15	-0.014	0.146	$2.021 \cdot 10^{-3}$
16	$-5.728 \cdot 10^{-3}$	0.095	$-5.469 \cdot 10^{-4}$
17	-0.018	0.054	$-9.81 \cdot 10^{-4}$
18	-0.013	0.024	$-3.25 \cdot 10^{-4}$
19	$2.511 \cdot 10^{-3}$	$6.156 \cdot 10^{-3}$	$1.546 \cdot 10^{-5}$
20	0.015	0	0

By comparing Figure 11.6 using the von Hann window coefficients to Figure 11.4 using Bartlett window coefficients, it is seen that the von Hann windowed FIR filter approaches the ideal lowpass filter more closely. The gain in Figure 11.6 is closer to 1 in the passband and has a narrower transition region. However, the computation of the von Hann coefficients is more complex. Usually the coefficients are precalculated and stored in memory, so the extra computation required for the von Hann coefficients is not important.

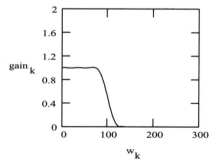

Figure 11.6. Von Hann windowed FIR lowpass filter gain for $N = 20$

Windows for FIR Filters

The Hamming Window

The Hamming window coefficients are computed using Equation 11.3. This equation is slightly more complex than the von Hann window coefficient equation. The only difference is that the constant and cosine terms are not both multiplied by 0.5 as in Equation 11.2, but rather each has its own multiplier. This makes the Hamming window a slight improvement over the von Hann window.

$$k_n = k_{-n} = 0.54 - 0.46 \cos\left[\pi * \frac{(N-n)}{N}\right] \quad \text{(Equation 11.3)}$$

As for the previous window coefficients, Equation 11.3 shows that they start at 0, go to 1, and then go to 0 again as n runs from $-N$ to 0 and then to N. Example 11.3 computes the Hamming window coefficients for the same ideal lowpass FIR filter used in the previous examples for $N = 20$. Figure 11.7 gives the gain curve of the Hamming windowed FIR lowpass filter.

Example 11.3. Calculating and using the Hamming window coefficients on a lowpass FIR filter

Problem: Given an ideal lowpass FIR filter with $w_p = 100$ rad/s, the sampling period $T = 0.01$s, and $N = 20$, compute the Hamming window coefficients and plot the windowed gain.

Solution:

$N := 20$

$n := 1 .. N$ The positive coefficient subscripts

$T := 0.01$ The sampling period in s

$w_p := 100$ rad/s   The lowpass frequency

$a_0 := w_p \cdot \dfrac{T}{\pi}$ The zero coefficient

$a_n := \dfrac{\sin(n \cdot w_p \cdot T)}{n \cdot \pi}$ The FIR coefficients

$k := 0 .. 100$ The number of data points to use to plot the gain

Digital Signal Processing

$w_k := 3 \cdot k$ The data point number converted to frequency in rad/s

$z_k := \cos(w_k \cdot T) + \sin(w_k \cdot T) \cdot j$

$hk_0 := 1$

$hk_n := 0.54 - 0.46 \cdot \cos\left(\pi \cdot \dfrac{N-n}{N}\right)$ The Hamming coefficients for positive n

$ah_0 := a_0 \cdot hk_0$ The noncausal Hamming windowed coefficient for n = 0

$ah_n := a_n \cdot hk_n$ The noncausal Hamming windowed coefficients for positive n

$Tz_k := ah_0 + \sum_n ah_n \cdot \left[(z_k)^n + (z_k)^{-n}\right]$

$gain_k := |Tz_k|$

n	a_n	hk_n	ah_n
1	0.268	0.994	0.266
2	0.145	0.977	0.141
3	0.015	0.95	0.014
4	-0.06	0.912	-0.055
5	-0.061	0.865	-0.053
6	-0.015	0.81	-0.012
7	0.03	0.749	0.022
8	0.039	0.682	0.027
9	0.015	0.612	$8.92 \cdot 10^{-3}$
10	-0.017	0.54	$-9.351 \cdot 10^{-3}$
11	-0.029	0.468	-0.014
12	-0.014	0.398	$-5.663 \cdot 10^{-3}$
13	0.01	0.331	$3.407 \cdot 10^{-3}$
14	0.023	0.27	$6.073 \cdot 10^{-3}$
15	0.014	0.215	$2.963 \cdot 10^{-3}$
16	$-5.728 \cdot 10^{-3}$	0.168	$-9.614 \cdot 10^{-4}$
17	-0.018	0.13	$-2.343 \cdot 10^{-3}$
18	-0.013	0.103	$-1.361 \cdot 10^{-3}$
19	$2.511 \cdot 10^{-3}$	0.086	$2.151 \cdot 10^{-4}$
20	0.015	0.08	$1.162 \cdot 10^{-3}$

Windows for FIR Filters

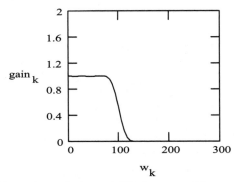

Figure 11.7. Hamming windowed FIR lowpass filter gain for $N = 20$

Figure 11.7 shows that the Hamming windowed gain is almost identical to the von Hann windowed gain for the lowpass filter used in all the examples of windowing. However, when a closer look is taken at the stopband by using gain in dB, it can be seen that there is a decrease in the stopband gain as the windowing goes from Bartlett to von Hann to Hamming in Figures 11.8a, 11.8b, and 11.8c, respectively. A corresponding reduction in the error in the passband could be seen if the previous plot were blown up around the passband gain of 1.

11.3 Non-Windowing Approaches

There are other ways to reduce the Gibbs effect than using windows, although windows are very popular and easy to use. Two other ways are mentioned here. There is nothing wrong with the coefficient integral equation developed in Chapter 10 for computing FIR coefficients. It was developed using correct mathematics, and it gives the correct values for each coefficient of an FIR filter given the ideal gain specifications for the four basic types of filters. The problem occurred because we are using only $2N + 1$ coefficients. In that case we have seen that modifying the coefficients with windowing produced better results in terms of ripple reduction versus widening of the transition region. One way around this problem is to make the coefficients converge faster by eliminating the jump included in the gain specifications for ideal filters; another is to mathematically find the coefficients that give equal ripple in the stopband. These two approaches are discussed next.

Digital Signal Processing

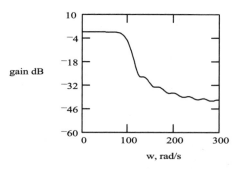

Figure 11.8a. Bartlett windowed FIR lowpass filter gain for $N = 20$

Jump Elimination in the Filter Specification

Since using only $2N + 1$ coefficients creates a transition region in the gain curve instead of the jump in the ideal specifications, one way to reduce the Gibbs effect is to specify the transition region anyway. This eliminates the jump in the ideal filter gain by including the specification of the transition region, which will appear anyway. Example 11.4 does this for the lowpass FIR filter used for all the previous examples in this chapter, with $N = 20$ again. The resulting gain curve is plotted in dB in Figure 11.9 along with the same filter gain with rectangular windowing (no windowing or correction for the Gibbs effect).

Example 11.4. Computing and using the FIR coefficients for a lowpass FIR filter with a transition region specified

Problem: Compute and use $N = 20$ FIR filter coefficients for a lowpass filter

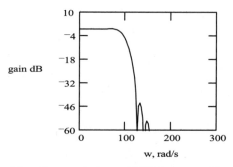

Figure 11.8b. Von Hann windowed FIR lowpass filter gain for $N = 20$

Windows for FIR Filters

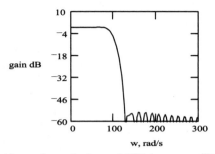

Figure 11.8c. Hamming windowed FIR lowpass filter gain for $N = 20$

with $w_p = 100$ rad/s and the sampling period $T = 0.01$. The transition region is specified to be 10 percent of the passband.

Solution:

$N := 20$ The number of positive subscripts of coefficients

$n := 0 .. N$ The positive coefficient subscripts

$T := 0.01$ The sampling period in s

$$a_n := \frac{1}{2 \cdot \pi} \cdot \left(\int_{-1}^{1} e^{j \cdot W \cdot n} \, dW \right)$$ The integral equation for coefficient calculation

$k := 0 .. 100$ The number of data points to use in plot

$w_k := 3 \cdot k$ The data point number converted to frequency in rad/s

$m := 1 .. N$

$$z_k := \cos(w_k \cdot T) + j \cdot \sin(w_k \cdot T)$$

$$Tz_k := a_0 + \sum_m a_m \cdot \left[(z_k)^m + (z_k)^{-m} \right]$$

$gain_k := |Tz_k|$ The gain using a rectangular window

$gaindB_k := 20 \cdot \log(gain_k)$

$$a1_n := \frac{-1}{2 \cdot \pi} \cdot \int_{1}^{1.1} (W - 1.1) \cdot 10 \cdot e^{j \cdot W \cdot n} \, dW$$

Digital Signal Processing

$$a2_n := \frac{1}{2\cdot\pi}\cdot\int_{-1.1}^{-1}(W+1.1)\cdot 10\cdot e^{j\cdot W\cdot n}\,dW$$

$$aa_n := a_n + a1_n + a2_n$$

$$Tz_k := aa_0 + \sum_m aa_m\cdot\left[(z_k)^m + (z_k)^{-m}\right]$$

$$gaina_k := |Tz_k|$$

$$gainadB_k := 20\cdot\log(gaina_k) \qquad \text{The gain with a specified transition region}$$

As can be seen in Figure 11.9, the specification of a transition region of 10 percent of the passband width has reduced the ripple in the stopband. However, an enlargement of the passband gain would show a slight increase in the passband ripple. Depending on the specifications, this method could be an improvement over the rectangular windowed FIR filter, but is not as good as the von Hann or Hamming windowed gain.

Parks-McClellan Method

The Parks-McClellan method does not use the integral equation developed in Chapter 10, which is mathematically correct. But since only $2N + 1$ coefficients are used, the resulting rectangular windowed approximation has ripple in the passband and stopband that is undesirable. One way to handle this ripple is to require it to have equal peaks

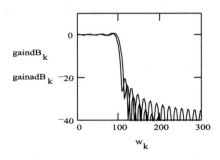

Figure 11.9. Comparison of lowpass filter gains with and without a transition region specified

throughout the passband and stopband. Unfortunately, equations to solve for this require knowledge of the frequencies at which the equal amplitude peaks occur. This leads to mathematics beyond the scope of this text. The problem is solved numerically by using the Remez exchange algorithm. Once it determines the frequency of the peaks, it is a relatively simple matter to solve for the coefficients that have equal amplitudes at these peaks. Figure 11.10 shows the gain in dB using the Parks-McClellan coefficients for the lowpass FIR filter with $N = 20$ used throughout this chapter. It should be pointed out that the Remez exchange algorithm works pretty well, but it is not guaranteed to find a solution. The Parks-McLellan coefficients can be obtained by using the Remez function in the Mathcad signals pack.

Application 2

Problem: Design a digital filter to be used to determine the relative magnitudes of signals in a certain frequency band for a stereo equalizer display. Only the design for the band between 500 and 2000 Hz, centered at 1000 Hz, will be shown. Once the digital filter is designed, the relative magnitude of the signals in the filter bandwidth can be approximated by

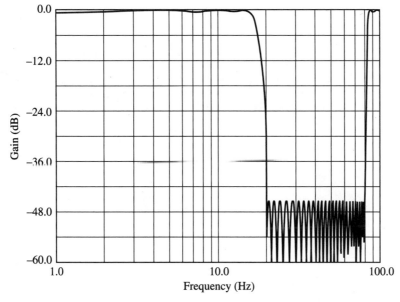

Figure 11.10. Parks-McLellan LP FIR, $N = 20$

digitally squaring the output samples, summing them over one cycle of the center frequency, and taking the square root of the result. This is the root-mean-square value or RMS value. This value can then be displayed by the processor as the height of a vertical bar, as is done on many stereos.

Solution: Let's try an FIR filter to select the sinusoidal components of the signal that have frequencies in the bandwidth of interest. Thus we need to first design (find the coefficients of) an ideal bandpass filter centered at 1 kHz. We know that by only using a finite number of ideal coefficients, we will produce the Gibbs effect as well as a finite transition band. So let's start with an ideal filter specification where the ideal passband lies between 750 Hz and 1500 Hz, since there will be a transition band due to the limited number of coefficients used.

In order to draw the ideal graphical filter specification, we need to select the sampling rate. Let's assume that this is a low-quality stereo and that the highest frequency in the signal is below 10 kHz. Then the Nyquist limit is satisfied if the sampling period is 0.00005 s, since this gives a sampling frequency of 20 kHz. Using the data given here, the ideal graphical specification is shown in Figure 11.11, since $\Omega = 2*\pi*f*T$.

Equation 10.6 gives the algebraic equation to use to compute the noncausal ideal FIR coefficients for a bandpass filter. Using this equation in Example 10.4 instead of Equation 10.4 for the lowpass filter coefficients, we get the gain plot in dB shown in Figure 11.12 for $N = 10$, or 21 coefficients. As this figure shows, the bandpass filter with $N = 10$ has a peak gain of about –2 db and is below –dB at 500 and 1500 Hz. This filter might be good enough, but there are peaks in the stopbands that are around –20 dB. If the signal had many components at these peak values at any time, a false reading of the relative magnitude might be obtained.

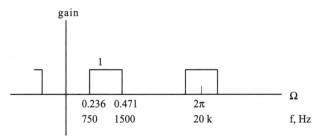

Figure 11.11. Ideal BPF graphical specification for Application 2

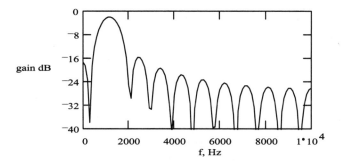

Figure 11-12. FIR lowpass filter gain for $N = 10$ for Application 2

To reduce this ripple in the stopbands, let's add windowing, which will further widen the transition regions but reduce the peaks of the ripple.

Example 10.4 is further modified by adding Hamming windowing as given in Equation 11.3. Figure 11.13 is a plot of the resulting gain in dB for $N = 10$ using Hamming windowing. As can be seen, the transition region widening is too great, so more coefficients are needed. By changing N to 15, we get the plot in Figure 11.14, which has better passband and stopband regions. The Mathcad program that plotted Figure 11.12 is run again with $N = 15$ to see if windowing is still needed if the N value is increased from 10 to 15. The gain plot in dB is shown in Figure 11.15, which shows little improvement in the ripple in the stopbands. Thus the Hamming windowed coefficients with $N = 15$ (31 coefficients) are used

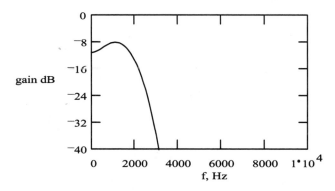

Figure 11.13. FIR lowpass filter gain for $N = 10$ for Application 2 with Hamming window

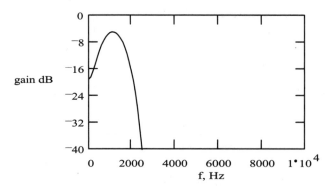

Figure 11.14. FIR lowpass filter gain for $N = 10$ for Application 2 with Hamming window

to implement the bandpass filter. The peaks of the ripple are less than −40 dB, so that even if significant signal amplitudes existed at these frequencies, their sum would not significantly change the RMS value. The widened bandwidth will include the effects of some signal amplitudes at frequencies beyond the desired passband, but it is better that the RMS value will be modified by the amplitudes of signal components near the center frequency than by those far from it.

In the following difference equation only a few windowed coefficients are shown since the FIR filter uses thirty-one coefficients. These coefficients can be obtained by simply typing ah[n= at the bottom of the modified Example 10.4 Mathcad program. This application example was used to show the implementation of an FIR filter in a simple application where the linear phase shift of the FIR filter is

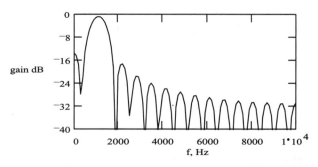

Figure 11.15. FIR lowpass filter gain for $N = 15$ for Application 2

Windows for FIR Filters

not needed. In a more sophisticated application of bandpass filtering, such as recombining the signals in each filtered band after a desired gain change is applied to each band, the nonlinear phase effects can cause significant audio effects that are not intended. Remember, a linear phase delay just means that all sinusoidal signal components are delayed the same amount in the time domain.

$$y(n) = 0.0019x(n+15) + \cdots + 0.075x(n) + \cdots + 0.0019x(n-15)$$

Finally, since the value displayed is the result of a current incoming signal whose RMS value of about 1 kHz is to be displayed, the causal form of the filter difference equation is needed, as shown next. A similar difference equation could be coded and used to replace the IIR filter difference equation in Application 1 in Chapter 7, as well as the equations to initialize and save previous input samples.

$$y(n) = 0.0019x(n) + \cdots + 0.075x(n-15) + \cdots 0.0019x(n-30)$$

Summary

This chapter has introduced the practical consideration of the effect of using only $2N + 1$ coefficients for an FIR filter, where N is the biggest positive subscript. For the ideal filter specifications, the jump at the passband edges leads to the Gibbs effect, which is a ripple in the passband and stopbands that approaches the jump frequency as N increases, but never goes away. This was shown in Figure 11.1. One solution is to decrease the magnitude of the coefficients as their subscripts approach N or $-N$. This approach is called windowing, and it reduces the ripple due to the jump from a passband gain of 1 to the stopband gain of 0. The price paid for this reduction is an increase in the transition bandwidth.

There are many different windowing methods; only a few were discussed in this chapter. The simplest computationally is the Bartlett window, which is just a linear reduction of the coefficients as their subscripts move away from zero. More-complex windowing schemes are the von Hann and the Hamming windows. They lead to a greater reduction in the ripple than the Bartlett for the same number of coefficients. Figures 11.4, 11.6, and 11.7 show the effects of these windowing schemes on the same lowpass FIR filter used in Figure 11.1, where $N = 20$ for all cases. Also, these last figures are repeated with the gain in dB in Figures 11.8a, 11.8b, and 11.8c. Plotting gain in dB better shows the ripple reduction in the stopband.

There are other ways to reduce the Gibbs effect. One is not to use ideal filter specifications to compute the FIR coefficients; that is, specify a finite width transition region. Figure 11.9 shows one example of this approach for the FIR filter in Figure 11.1. Another method is not to use the FIR filter coefficient equation developed in Chapter 10 at all. The Parks-McClellan method mathematically finds the coefficients that would yield the minimum amplitude of the ripple if it were evenly distributed throughout the passband and stopband. The solution is a minimax problem that uses the Remez exchange procedure, which is beyond the scope of this text. However, Figure 11.10 shows the results of using the Parks-McClellan method on the lowpass filter with $N = 20$ that was plotted in Figure 11.1. The equal ripple of this method is sometimes preferred, but it does not always produce the lowest ripple overall.

Self-Test

1. Compute the noncausal Bartlett window coefficient for the noncausal a_3 FIR coefficient where the maximum positive subscript of the filter is $N = 10$.

2. Compute the noncausal von Hann window coefficient for the noncausal a_{-5} FIR coefficient where the maximum positive subscript of the filter is $N = 20$.

3. Compute the noncausal Hamming window coefficient for the noncausal a_{10} FIR coefficient where the maximum positive subscript of the filter is $N = 15$.

4. Compute the causal Bartlett window coefficient for the causal a_7 FIR coefficient where the maximum positive causal subscript is $2N = 10$.

5. Compute the causal von Hann window coefficient for the causal a_1 FIR coefficient where the maximum positive causal subscript is $2N = 8$.

6. Compute the causal Hamming window coefficient for the causal a_9 FIR coefficient where the maximum positive causal subscript is $2N = 18$.

7. Use Mathcad to compute the nine noncausal Bartlett window coefficients for $N = 4$.

Windows for FIR Filters

8. Use Mathcad to compute the eleven noncausal von Hann window coefficients for $N = 5$.

9. Use Mathcad to compute the fifteen noncausal Hamming window coefficients for $N = 7$.

Problems

1. Compute the noncausal Bartlett window coefficients for the noncausal a_{-2} FIR coefficient where the maximum positive subscript of the filter is $N = 5$.

2. Compute the noncausal von Hann window coefficient for the noncausal a_3 FIR coefficient where the maximum subscript of the filter is $N = 10$.

3. Compute the noncausal Hamming window coefficient for the noncausal a_5 FIR coefficient where the maximum positive subscript of the filter is $N = 7$.

4. Compute the causal Bartlett window coefficient for the causal a_4 FIR coefficient where the maximum positive subscript is $2N = 12$.

5. Compute the causal von Hann window coefficient for the causal a_7 FIR coefficient where the maximum positive causal subscript is $2N = 10$.

6. Compute the causal Hamming window coefficient for the causal a_2 FIR coefficient where the maximum positive causal subscript is $2N = 16$.

Answers to Self-Test

1. $k_3 = 0.7$

2. $k_{-5} = 0.854$

3. $k_{10} = 0.310$

4. k_7 (causal) $= 0.6$

5. k_1 (causal) = 0.146

6. k_9 (causal) = 1

7. $k_4 = 0.0$, $k_{-2} = 0.5$, $k_0 = 1$

8. $k_{-5} = 0.0$, $k_{-3} = 0.345$, $k_0 = 1.0$

9. $k_{-7} = 0.08$, $k_{-4} = 0.438$, $k_0 = 1.0$

chapter 12

Practical Digital Filter Considerations

Introduction

This chapter covers some of the practical problems and solutions related to implementing digital filters. The first consideration in using a digital filter is to decide whether an IIR or an FIR filter is to be used. Because of their distinct characteristics, each is used almost exclusively in certain sections of the industry. In this chapter we briefly discuss the advantages and disadvantages of each type of filter and show why certain segments of the industry prefer each type.

In order to reduce the effects of numerical quantization, digital IIR filters of any order are usually implemented as combinations of first- and second-order types. These first- and second-order filters may be combined in parallel or in series. Also, for IIR filters, a filter of any order can be mathematically manipulated into a form requiring fewer steps and less memory; the simplest form is called the **canonical form**.

In Chapter 9 we discussed the effect of numerical quantization of the filter coefficients on the resulting filter, but the numerical quantization of the signal needs to be considered here. This leads to the determination of the number of bits used by the ADC and the DAC. Finally, the choice between encoding the samples as fixed point or floating point numbers is discussed, and the basic effects of either choice are pointed out.

Digital Signal Processing

12.1 FIR versus IIR Digital Filters

The choice of using either an IIR or an FIR filter affects almost all the other design criteria. Many segments of the industry use either one or the other exclusively, which may make the choice easier, but the designer needs to know what advantages or disadvantages have been given up. Also, some filter applications may be solved by either type, and then the following items must be taken into account at the outset.

The major difference between an FIR and an IIR filter is that the first uses only the a coefficients, while the latter uses both a and b coefficients. On the surface then, an FIR may seem like an obvious choice for simplicity of design and implementation. But you pay a price for using only the a coefficients. You usually don't get something for nothing, which holds true here. By selecting an FIR filter using only the a coefficients, you usually require many more coefficients than the corresponding IIR filter, if there is a corresponding IIR filter. This can be seen in Figures 12.1, 12.2, 12.3, and 12.4.

Figure 12.1 shows the gain plot of an FIR lowpass filter with $w_p = 50$ rad/s and the sampling period $T = 0.01$ seconds. This gives $\Omega_p = 0.5$. The FIR filter coefficients are computed using Equations 10.4a and 10.4b, where the number of coefficients used is 9, or $N = 4$. Figure 12.2 is the gain plot of an IIR Butterworth filter developed using the BLT method to meet the specifications of the FIR filter in Figure 12.1, which is down 3 dB at 39 rad/s and below −18dB at 75 rad/s. This IIR filter uses seven coefficients to meet the same specifications, which does not seem like much of a reduction in coefficients. The transfer function for this IIR filter is given in the following equation.

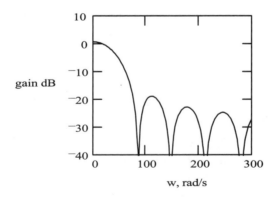

Figure 12.1. FIR lowpass filter gain for $N = 4$

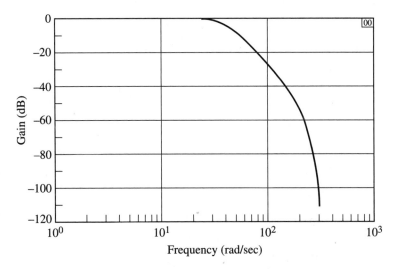

Figure 12.2 IIR Butterworth filter gain using BLT method

$$H(z) = \frac{0.005032(z+1)^3}{(z-0.6736)[(z-0.7801)^2 + 0.2739^2]}$$

However, let's tighten the specifications a little more. From Figure 12.2 we can see that the IIR filter actually has no stopband ripple, since it is a Butterworth filter. Let's require the FIR filter to have stopband ripple below −40 dB. Figure 12.3 shows the rectangular windowed (this really means no windowing) FIR lowpass filter with $N = 20$, or 41 coefficients. It is seen that even using forty-one coefficients the requirement that the

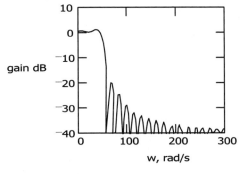

Figure 12.3. FIR lowpass filter gain for $N = 20$

stopband ripple be under −40 dB can't be met. Using the Hamming window coefficients on the lowpass filter coefficients with $N = 6$, or 13 windowed coefficients, gives the gain plot in dB in Figure 12.4. Measurements from Figure 12.4 show that the gain is −3 dB at 27 rad/s and reduces below −40 dB above 130 rad/s. This is approximately what the IIR butterworth filter gain does in Figure 12.2, but now thirteen coefficients are used instead of seven for the gain shown in Figure 12.2.

The increase in the number of coefficients from using an FIR filter instead of an IIR filter means that more memory and computations are required; the computations reduce the maximum sampling period T that can be used. Also the use of more a coefficients, three for the IIR filter in Figure 12.2 and thirteen for the FIR filter plotted in Figure 12.4, means a longer delay between input and output, since you can't compute any output until all the required input samples have been obtained. This delay is not usually critical, unless real-time digital signal processing is required. For real-time signal processing this extra delay caused by the FIR filter can be very significant.

The preceding discussion about time delay brings up an advantage of using an FIR filter. This advantage is that there is no phase shift for a noncausal FIR filter using the integral equation in Chapter 10 and a linear phase delay for a causal FIR filter (one used for real-time processing). This is because the development of the integral equation in Chapter 10 used the magnitude only in its derivation; thus the phase shift was assumed to be zero. In order to process in real time, the noncausal coefficients must be converted to the causal coefficients. Remember, this is just a renaming so that the computation can proceed; it really amounts to delaying the output until the required input samples are available.

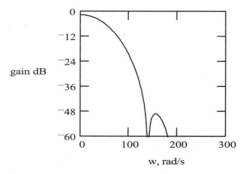

Figure 12.4. Hamming windowed FIR lowpass filter gain for $N = 6$

Practical Digital Filter Considerations

Students can recall from AC circuits class that a delay is a linear phase shift of all the frequencies in the signal. However, since the IIR filter is an approximation to an analog filter, the phase shift of the filter is not controllable. Students should recall from analog signal processing classes that a first-order factor goes through a 90-degree phase shift, and second-order filters go through 180 degrees of phase shift.

Besides being able to be designed with linear phase shift, or a constant delay without distortion, FIR filters also have no stability problems. Remember, there are no b coefficients and thus no denominator terms in the transfer function, $T(z)$, of FIR filters. This is shown in the following expression, where Equation 10.2 is now given as Equation 12.1.

$$T(z) = a_0 + a_1 z^{-1} + \cdots + a_N z^{-N} + a_{-1} z + \cdots + z^N a_{-N} \quad \text{(Equation 12.1)}$$

Since Equation 12.1 shows that there are no poles of $T(z)$, it is impossible to design an unstable FIR filter or to cause instability by using too low a coefficient precision. Sometimes this stability property, along with the linear phase shift produced by FIR filters, is more important than the effects of usually using more coefficients than IIR filters. The filter designer must be aware of these effects and make the choice based on the applications and specifications required.

12.2 Effects of Analog to Digital Converter Number of Bits

The number of bits used to convert the analog signal to a sampled time signal affects the **Signal to Noise Ratio** or **SNR** of the signal sent into the filter. The SNR is the ratio of the signal amplitude to the unwanted signal or noise amplitude introduced into the signal by using a limited number of bits to represent the sampled signal amplitude. The SNR is usually expressed in dB, just as the gain of the filter is, which is another ratio. The noise in the SNR is called **quantization noise**, because the limited number of bits quantizes the sampled signal into discrete magnitude values. This effect is shown in Figure 12.5.

In order to quantitatively determine the SNR in terms of the number of ADC bits used, the mathematics of stochastic processes would have to be used. But we will present here a logical derivation of the relationship that agrees with the results of using the mathematics of stochastic processes. The following steps should allow the student to more fully understand the

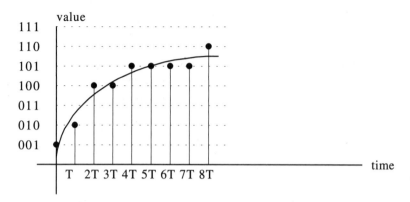

Figure 12.5. Quantized sample values of a signal using unsigned magnitude with three bits

relationship between input noise and the number of ADC bits used, as well as to modify the results to their special cases.

Let's first look at the best we can do to represent an analog signal into an ADC using only one bit. Let us restrict the signal so that its amplitude varies between 0 and 1. A low signal level will be represented with a 0 value and a high level with a 1. If the one-bit ADC symmetrically rounds the sampled value of the analog signal, then any sample of the input signal over 0.5 would be represented by the ADC with a 1, and lower signal levels with a 0. This is shown in Figure 12.6, where the maximum error would be 0.5, with linearly decreasing errors for other sample values.

Now the maximum error to the one-bit ADC is 0.5 for a signal that has values between 0 and 1. But statistically that maximum error is not reached too often. The method used to statistically represent the quantization error is to find the one sigma or standard deviation of the error. The mathematics is a little beyond the scope of this text, since the error is uniformly distributed instead of having a Gaussian distribution, as can be seen from Figure 12.6. In Figure 12.6, if any signal value is equally likely, then any error is equally likely; this is called a *uniform distribution*. Using the methods of stochastic processes, the standard deviation of a uniformly distributed error is twice the maximum error divided by the square root of twelve. This equation is given by Equation 12.2.

Practical Digital Filter Considerations

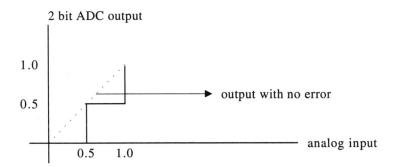

Figure 12.6. Magnitude of quantization and error using one bit

$$\sigma = \frac{2 * \text{max\_error}}{\sqrt{12}} \quad \text{(Equation 12.2)}$$

The one sigma error given by Equation 12.2 is 0.289 or −10.8 dB. This seems reasonable when looking at Figure 12.6. There the expected error would be between 0 and 0.5.

Now let's look at a reasonable statistical representation of the signal. We have said that it varies between 0 and 1 in value, so any sample of it by the one-bit ADC can be any value in between. A reasonable expected value would be 0.5 or −6 dB. Taking the ratio of the two statistical representations gives a crude SNR of a one-bit ADC. This ratio is 1.74 or 4.8 dB, where the SNR in dB is just the signal in dB minus the noise in dB.

Let's consider the effect of using more than one bit for the ADC. From Figure 12.7 it can be seen that using 2 bits to quantify the sampled input signal reduces the maximum error to 0.25 or half the error of a one-bit ADC. All the other quantization errors besides the maximum are also reduced by the same amount. Thus, it is reasonable, that every bit added to an ADC reduces the quantization noise by half or −6 dB. This gives the final quantization noise or error due to using B number of bits in an ADC as Equation 12.3, where σ, the one sigma value of the SNR, is in dB.

$$\sigma = 6(B-1) + 4.8 \quad \text{(Equation 12.3)}$$

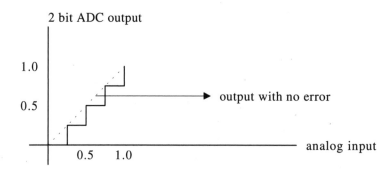

Figure 12.7. Magnitude quantization and error using two bits

Example 12.1. Computing the SNR for an 8 bit ADC

Problem: Let the ADC that is sampling the input signal into a digital filter use eight bits to represent the value of the sampled signal at each sample time.

Solution: The number of bits is B = 8. Using Equation 12.3, the SNR = 46.8 dB.

12.3 Fixed Point Math versus Floating Point for a DSP Chip

General purpose computers and microprocessors do their numerical calculations using floating point 2's complement arithmetic. In other words, the numbers consist of a fractional part called the mantissa and an exponential part, each coded in 2's complement arithmetic. The exponential part just tells how many places to the right or left the binary digit should be shifted for the fractional part to represent the true value represented. Thus the number of bits representing the precision of the value represented does not change. This is called the **floating point** number representation. This scheme maintains the same precision for any value coded by the number, but the scaling by the exponent requires extra time when two numbers are added, subtracted, multiplied, or divided. It is easier just to eliminate the exponent and use all the bits to represent the number. This is called **fixed point** number representation. However, this limits the range of values that the number can represent, since the binary point is fixed. Because of the increase in speed using this fixed point representation, many DSP chips use fixed point arithmetic. Examples 12.2 and 12.3 give the limits of the values that can be represented by fixed point and floating point numbers using six bits.

Practical Digital Filter Considerations

Example 12.2. Computing the range of values represented by a fixed point number

Problem: Let the number of bits used be 6, and let the binary numbers be represented by 2's complement arithmetic.

Solution: The maximum positive number is binary 011111, which is decimal 31, since each bit position represents a multiplication of the number in the position by 2 as one goes left. The left-most bit just says the number is positive. The minimum negative number is binary 111111, which is decimal −32, since the magnitude of the number is given by inverting all the bits and adding 1. The smallest nonzero numbers represented are 1 and −1.

Example 12.3. Computing the range of values represented by a floating point number

Problem: Let any value be encoded using three bits for a fractional part in 2's complement form with the exponent using three bits in unsigned form.

Solution: In fractional form each bit position to the right of the sign bit is half of the preceding bit, with the first bit to the right of the sign bit representing one half. Thus the fractional part ranges from 0.75 to −1 in steps of 0.25. However, the exponent tells how many bit positions to the right to move the binary point, which for three bits in unsigned form can range between 0 and 7. Thus the maximum range of values represented goes from −128 to 96, with the smallest nonzero values still being −1 and 1.

From Examples 12.2 and 12.3 it is seen that a wider range of numbers can be represented using floating point numbers with the smallest numbers still the same. This is true in general for floating point numbers and is the reason they are used almost exclusively in general-purpose computers and microprocessors. However, floating point computation takes longer since when adding, subtracting, multiplying or dividing, more steps are needed to put the result in the form of a fractional mantissa and compute the correct exponent for the fractional mantissa. Thus when speed is paramount, DSP chips use fixed point numbers. However, with a smaller range of values represented, care must be taken to avoid numeric overflow when mathematical operations are performed. Great care must be taken so that when overflow occurs, the maximum value is held and rollover does not occur.

12.4 Realization Forms for Filters

The difference equations representing IIR filters can be manipulated to represent the filter using various mathematical operations giving the same result. This gives different realizations of IIR filters when they are coded. Four different realizations are given in this section, the first being the direct form using the standard notation for a recursive difference equation. However, some simple mathematical manipulation gives the canonical form, which uses less storage of previous values. Finally, the denominator of the transfer function for IIR filters is usually factored into first- and second-order factors to reduce the effect of numerical precision on the filter. The IIR filter transfer function can then be written as a product of numerator factors over denominator factors, or the transfer function can be written as a partial fraction expansion as given in any analog signal processing course or algebra course. Each of these product terms or sum terms can be in direct or canonical form.

The Direct Form of the IIR Filter

The **direct form** of the IIR filter is obtained by coding the difference equation directly as it is usually written, as shown in Equation 5.2. Equation 5.2 is repeated here as Equation 12.4.

$$y(n) = a_0 x(n) + a_1 x(n-1) + \cdots + a_N x(n-N)$$
$$+ b_1 y(n-1) + b_2 y(n-2) + \cdots + b_M y(n-M) \quad \text{(Equation 12.4)}$$

Using Equation 12.4, the graphical representation of the direct form realization is given in Figure 12.8 for $M = 2$ and $N = 3$. Figure 12.8 is written using the z-transform of the signals so that the transfer function for a time delay of one sample period, z^{-1}, can be used. The symbols for the inverse z-transforms of the input and output are written below their z-transforms. It is easy to see that the output signal in Figure 12.8 is given by Equation 12.4. Figure 12.8 shows that the direct form of coding an IIR filter with $M = 2$ and $N = 3$ requires five delays, six multiplies, and five summations. The delays are done by storing the values in some type of computer or DSP memory for use one cycle later. Now let's look at this same IIR filter using the canonical form for coding.

Practical Digital Filter Considerations

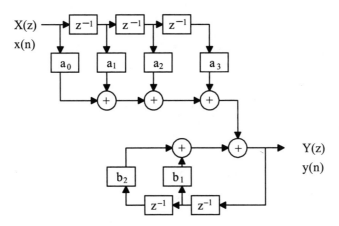

Figure 12.8. Direct realization of an IIR filter with four *a* coefficients and two *b* coefficients

The Canonical Form of the IIR Filter

The **canonical form** or realization is coded using Equations 12.5a and 12.5b.

$$Y(z) = W(z) \sum_{k=0}^{N} a_k z^{-k} \qquad \text{(Equation 12.5a)}$$

$$W(z) = X(z) + W(z) \sum_{k=1}^{M} b_k z^{-k} \qquad \text{(Equation 12.5b)}$$

It is shown next that Equations 12.5a and 12.5b give the z-transform of Equation 12.4. The mathematical steps for this start by using Equation 12.5b. The purpose for introducing the signal $W(z)$ will be shown by looking at the graphical representation in Figure 12.9.

$$W(z) - W(z) \sum_{k=1}^{M} b_k z^{-k} = X(z)$$

$$W(z) * \left[1 - \sum_{k=1}^{M} b_k z^{-k} \right] = X(z)$$

$$W(z) = \frac{X(z)}{1 - \sum_{k=1}^{M} b_k z^{-k}}$$

$$Y(z) = \left[\frac{X(z)}{1 - \sum_{k=1}^{M} b_k z^{-k}} \right] * \sum_{k=0}^{N} a_k z^{-k}$$

$$Y(z) \left[1 - \sum_{k=1}^{M} b_k z^{-k} \right] = X(z) * \sum_{k=0}^{N} a_k z^{-k}$$

$$Y(z) = Y(z) \sum_{k=1}^{M} b_k z^{-k} + X(z) \sum_{k=0}^{N} a_k z^{-k}$$

$$Y(z) = \sum_{k=1}^{M} b_k z^{-k} Y(z) + \sum_{k=0}^{N} a_k z^{-k} X(z)$$

Taking the inverse z-transform of the last equation using the shifting property gives Equation 12.4. Now let's look at the graphical representation of a digital IIR filter coded using Equations 12.5a and 12.5b with $M = 2$ and $N = 3$. This is shown in Figure 12.9. Figure 12.9 shows that the IIR filter with $M = 3$ and $N = 2$ coded using the canonical realization uses three delays, six multiplies, and five additions. Compared with the direct

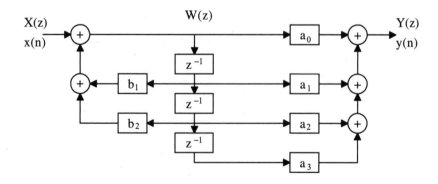

Figure 12.9. Canonical realization of an IIR filter with four *a* coefficients and two *b* coefficients

Practical Digital Filter Considerations

form of realization, the canonical form uses less delays. The delays are memory locations, since a value is delayed by storing it and using the value in the next computer or DSP cycle.

The Cascade Form of the IIR Filter

As previously shown, the IIR filter can be realized or coded in direct form or canonical form. This is taken even further by factoring the numerator and denominator of the transfer function $T(z)$ of the IIR filter into first- and second-order factors. Then $T(z)$ could be written as a product of terms with first- and second-order numerators and denominators. This is usually done, since it contributes to numerical stability. A simple example is a pole or denominator factor of $(z - 0.987)$; no amount of truncation or rounding will cause the pole to have a magnitude greater than one, which would make the entire filter unstable. However if several poles are multiplied together to form the denominator, numerical truncation or rounding of the coefficients of the resulting polynomial in z could produce a pole magnitude greater than one. The terms are usually then written in the canonical first- or second-order form. These terms are then each converted to difference equations using the shifting property, and each difference equation is then coded. This **cascade form** is illustrated in Example 12.4.

Example 12.4. Determining the cascade form using direct form terms

Problem: Given the factored transfer function $T(z)$ of an IIR filter below, determine a cascade form each written as a first- or second-order direct form.

$$T(z) = \frac{0.8(z-1)(z+0.9)(z-0.9)}{(z+0.8)(z^2-1.2z+0.85)}$$

Solution: One solution would be to group the numerator and denominator factor as shown in the following equation, and then determine the corresponding difference equations as shown.

$$T(z) = \left[\frac{0.8(z-1)}{z+0.8}\right]\left[\frac{z^2-0.81}{z^2-1.2z+0.85}\right]$$

Let's call the first term on the right $T_1(z)$ and the second term $T_2(z)$. Then the overall filter transfer function could be implemented by coding the following two equations, which are obtained in the usual way from $T_1(z)$ and $T_2(z)$. If the cascade filter is implemented with the term representing $T_1(z)$ before the other term, then $x_1(n)$ is really the overall input $x(n)$, $y_2(n)$ is the overall output $y(n)$, and the output of the first filter $y_1(n)$ is the input to the second filter $x_2(n)$. This leads to the last two equations, which would then be coded to give a cascade form of the original filter.

$$y_1(n) = -0.8y_1(n-1) + 0.8x_1(n) - 0.8x_1(n-1)$$

$$y_2(n) = 1.2y_2(n-1) - 0.85y_2(n-2) + x_2(n) - 0.81x_2(n-2)$$

Or:

$$y_1(n) = -0.8y_1(n-1) + 0.8x(n) - 0.8x(-1)$$

$$y(n) = 1.2y(n-1) - 0.85y(n-2) + y_1(n) - 0.81y_1(n-2)$$

The Parallel Form of the IIR Filter

If only the denominator of the transfer function of an IIR filter is factored into first- and second-order terms, the **Partial Fraction Expansion** (PFE) method from algebra can be used to write the transfer function $T(z)$ as a sum of terms with first- and second-order denominators. Again, by having only first- and second-order denominators, it is less likely that numerical truncation or rounding will cause instability. Remember that PFE is just the inverse mathematical operation of putting a sum of rational terms over a common denominator. However, using PFE to obtain the original transfer function as a sum of terms is tricky for z-transforms. This is due to the fact that the numerator order must be less than the denominator order in order to directly use PFE. But many $T(z)$ have the same order numerators as denominators, so some algebraic trickery is used, which will not be gone into here.

The Transversal Form for FIR Filters

The FIR filter is usually coded directly from the difference equation of the FIR filter. This is called the **Transversal form**, since the filter operations proceed transversely along a line in the graphical representation for the transversal form of the FIR filter. Equation 12.1 gives the causal difference equation for an FIR filter, so it is not repeated here. Figure 12.10 gives the graphical representation of the transversal form of the FIR filter. Remember that an FIR filter can be causal or noncausal.

Summary

In this chapter we have looked at some of the practical aspects of using digital filters. The first was how to choose between IIR and FIR filters. Some of the characteristics of each were gone over. The primary characteristics of IIR filters is that they are digital approximations to analog filters. They will have the same phase characteristics and stability concerns as do analog filters. By using the integral equation in Chapter 10, FIR coefficients can be found for any specified phase shift. In Chapter 10, the integral equation was solved to give the coefficients for ideal gain specifications for the four basic types of filters without any phase shift.

These algebraic solutions were for noncausal filters, which require input samples for the current output that have not occurred yet. But this is not a problem if the input samples are already stored in a computer. The causal form is obtained by renaming the coefficients, which essentially delays the output until all the input samples required are available. This delay causes a linear phase shift with respect to the input frequency if the noncausal filter had no phase shift. The resulting delay may be a problem in real-time or control system applications. However, the major problem with FIR filters is that they usually require more coefficients than an IIR filter. This was illustrated in Figures 12.1 through 12.4.

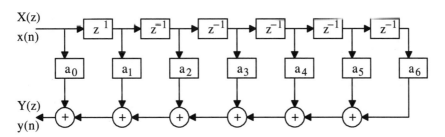

Figure 12.10. The Transversal realization of a seven-coefficient causal FIR filter

Another concern to the digital filter designer is the magnitude quantization problem of the input samples due to the number of bits used in the DAC. A simple equation, Equation 12.3, was developed to compute the approximate standard deviation of this quantization noise in order to determine its effect on the filter specifications.

The basic problem of using floating point numbers versus fixed point numbers was discussed and their characteristics given. Mathematical operations with floating point numbers are slower, but overflow or numeric rollover is less likely than with fixed point numbers. The filter designer must weigh these characteristics against the filter requirements to determine which method is best.

Finally the problem of coding the filter was covered. By mathematical manipulation the difference equations to be coded can be written in various ways, called forms, implementations, or realizations. For IIR filters, directly coding the difference equation led to the direct form. It was shown that the difference equation can always be rewritten so that less memory is required. This led to the canonical form, which is almost always used. Higher-order IIR filter transfer functions can be factored into product of terms, and the difference equations for each can be coded directly or using the canonical form. By using product of terms with first- or second-order factors, it is impossible to cause IIR filter instability due to coefficient truncation or rounding. The transfer function of the IIR filter could also be represented by a sum of terms using PFE, each term also having a first- or second-order denominator. This parallel realization was not pursued here, since special techniques are required to PFE z-transfer functions.

Self-Test

1. Determine the approximate SNR due to magnitude quantization when twelve bits are used for the ADC.

2. Determine the approximate SNR due to magnitude quantization when sixteen bits are used for the ADC.

3. Draw the diagram for the direct form of the realization for the following digital filter transfer function, similar to that shown in Figure 12.8.

$$T(z) = \frac{0.2z}{z - 0.817}$$

Practical Digital Filter Considerations

4. Given the following transfer function of a digital IIR filter $T(z)$, write the two second-order transfer functions $T_1(z)$ and $T_2(z)$ that could be cascaded to have the same transfer function as $T(z)$, and then give the difference equations of each with each one having the same a_0 value.

$$T(z) = \frac{5.6415 * 10^{-4}(z+1)(z+1)(z+1)(z+1)}{(z^2 - 1.7171z + 0.9225)(z^2 - 1.7717z + 0.8156)}$$

5. Use the canonical form to determine the difference equations for the following first-order transfer function.

$$T(z) = \frac{z}{z - 0.368}$$

6. Write the z-transformed equations for the canonical form for the filter with the following transfer function, and then give the corresponding difference equations.

$$T(z) = \frac{0.1277(z+1)(z+1)}{z^2 - 0.7664z + 0.2774}$$

7. Draw the canonical realization of the filter $T_1(z)$ in Problem 4 similar to the realization shown in Figure 12.9.

8. Draw the canonical realization of the filter in Problem 5 similar to the realization shown in Figure 12.9.

Problems

1. Determine the approximate SNR due to magnitude quantization when eight bits are used for the ADC.

2. Determine the approximate SNR due to magnitude quantization when four bits are used for the ADC.

3. Draw the diagram for the direct form of the realization for the following digital filter transfer function, similar to that shown in Figure 12.8.

$$T(z) = \frac{0.8(z-1)}{z - 0.6}$$

Digital Signal Processing

4. Given the following transfer function of a digital IIR filter $T(z)$, write the two first-order transfer functions $T_1(z)$ and $T_2(z)$ that could be cascaded to have the same transfer function as $T(z)$, and then give the difference equations of each one having the same a_0 value.

$$T(z) = \frac{0.2097z^2}{(z^2 - 0.75z + 0.243)(z^2 - 1.03z + 0.525)}$$

5. Use the canonical form to determine the difference equations for the following first-order transfer function.

$$T(z) = \frac{0.632}{z - 0.638}$$

6. Write the z-transformed equations for the canonical form for the filter with the following transfer function, and then give the corresponding difference equations.

$$T(z) = \frac{0.172z}{z^2 - 1.32z + 0.49}$$

7. Draw the canonical realization of the filter $T_1(z)$ in Problem 4 similar to the realization shown in Figure 12.9.

8. Draw the canonical realization of the filter in Problem 5 similar to the realization shown in Figure 12.9.

Answers to Self-Test

1. 71 dB

2. 95 dB

3. $a_1 = 0.2$
   $b_1 = 0.817$

4. $T_1(z) = \dfrac{0.02375(z+1)(z+1)}{z^2 - 1.7171z + 0.9225}$

   $T_2(z) = \dfrac{0.02375(z+1)(z+1)}{z^2 - 1.7717z + 0.8156}$

Practical Digital Filter Considerations

$y_1(n) = 1.7171y_1(n-1) - 0.9225y_1(n-2) + 0.02375[x_1(n) + 2x_1(n-1) + x_1(n-2)]$

$y_2(n) = 1.7717y_2(n-1) - 0.8156y_2(n-2) + 0.02375[x_2(n) + 2x_2(n-1) + x_2(n-2)]$

5. $w(n) = x(n) + 0.368w(n-1)$
   $y(n) = w(n)$

6. $Y(z) = W(z)[0.1277(1 + 2z^{-1} + z^{-2})]$
   $W(z) = X(z) + W(z)[0.7664z^{-1} - 0.2774z^{-2}]$
   $y(n) = 0.1277[w(n) + 2w(n-1) + w(n-2)]$
   $w(n) = x(n) + 0.7664w(n-1) - 0.2774w(n-2)$

7. $a_0 = 0.0237 \quad a_1 = 0.0475 \quad a_2 = 0.0237$
   $b_1 = 1.717 \quad b_2 = -0.9225$

8. $a_1 = 1.0$
   $b_1 = 0.368$

chapter 13

Digital Integration

Introduction

In this chapter we take a very brief look at using the computer to perform integration, called **digital integration**. This is a very complex subject, which comprises thousands of times more material than can be discussed here. The purpose of this chapter is to show another use for DSP besides digital filtering and to illustrate some of the basic uses and characteristics of digital integration. We only look at very simple digital integration, just so the student has an idea of the subject and characteristics. The techniques given in this chapter are not the best or even the usual methods for digital integration, but they are useful for illustrative purposes, since they are simple and easily understood, and are chosen for that reason.

13.1 Introduction to Digital Integration

There are three fundamental uses for digital integration. The first use is to integrate a set of samples of signals or data that are all known and stored values. In this use, Simpson's Rule or the trapezoidal rule given in algebra or calculus texts is used. The trapezoidal rule is given in Equation 13.1, and Simpson's rule is in Equation 13.2.

$$\int_0^{nT} x(t)dt \approx T[0.5x(0) + x(1) + \cdots + x(n-1) + 0.5x(n)]$$

(Equation 13.1)

$$\int_0^{nT} x(t)\,dt \approx \frac{T}{3}[x(0) + 4x(1) + 2x(2) + 4x(3) + 2x(2) + \cdots$$

$$+ 4x(n-1) + x(n)] \qquad \text{(Equation 13.2)}$$

The second use is to integrate signals or samples of signals in real time, that is, to compute the current integral before the next input sample is taken. This method and its requirements will be illustrated by modification of the trapezoidal rule.

The final use of digital integration is to find the solution of differential equations, which is of primary concern in engineering analysis and simulation. In this chapter we will use very basic digital integration procedures, so that the concepts can be understood. Let's look at a simple example of integrating known samples of data using the trapezoidal rule, in Example 13.1.

Example 13.1. Using the trapezoidal rule to find the integral of sampled data

Problem: Given the following uniformly sampled values of a signal, find the area under the curve using the trapezoidal rule. The samples are separated by $T = 0.15$ s.

$x(0) = 0.2 \qquad x(3) = 0.3 \qquad x(6) = 0.4$

$x(1) = 0.4 \qquad x(4) = 0.7 \qquad x(7) = 0.3$

$x(2) = 0.5 \qquad x(5) = 0.6 \qquad x(8) = 0.5$

Solution: Using Equation 13.1, with $n = 8$, we have the following equation.

$$y(8T) = y(8) = 0.15[0.5(0.2) + 0.4 + 0.5 + 0.3 + 0.7 + 0.6 + 0.4 + 0.3 + 0.5(0.5)] = 0.5325$$

Given only the samples of the signal that is to be integrated, there is no way to know if the trapezoidal rule gives the correct answer in Example 13.1. All that can be said is that Simpson's rule is usually more accurate. The most important point to be made about Example 13.1 is that only the total area under the sampled signal at the end of the sampling is given. In many engineering applications the integral of the current signal

Digital Integration

is needed at each sample time, such as determining current position by integrating the output of a tachometer. The trapezoidal rule or Simpson's rule could be repeated for every new input sample from the ADC, but repetition is time-consuming. The next section shows how the trapezoidal rule can be broken down to allow the current integral to be computed.

13.2 Digital Integration of Known Signals

If a signal is being sent through an ADC so that digital integration can be performed on it, Simpson's rule or the trapezoidal rule can be simplified so that fewer calculations are needed. Note from Equation 13.1 that the first and last samples must be divided by 2. This is no problem for the initial sample since once it is done, it is done. But if the signal continues to be sampled, the last (or most current sample) value keeps changing. So if the trapezoidal rule is used to integrate a signal continuously as samples of it are taken, you can't just add half of it to the previous integral, since it used half of the previous sample value and now you need its full value and half the current sample value. The problem is even more complex for Simpson's rule, given in Equation 13.2.

Let's look at how the trapezoidal rule was developed and modify it so that it can be used on a signal that is continuously being sampled and digitally integrated. Figure 13.1 illustrates how the trapezoidal rule approximates the area under a signal. In Figure 13.1, the area under the smooth curve is $y(t)$, and it is approximated by summing the area of all the trapezoids whose upper vertices are on the curve. It can be seen that this sum approximates the area under the smooth curve, especially if the time between samples decreases. Let's sum the trapezoidal areas to derive the trapezoidal rule as shown in the following equations.

$$y(t) \approx T\frac{x(0)+x(1)}{2} + T\frac{x(1)+x(2)}{2} + \cdots + T\frac{x(n-1)+x(n)}{2}$$

$$y(t) \approx T[\frac{x(0)}{2} + \frac{x(1)}{2} + \frac{x(1)}{2} + \frac{x(2)}{2} + \cdots + \frac{x(n-2)}{z} + \frac{x(n-1)}{2} + \frac{x(n-1)}{2} + \frac{x(n)}{2}]$$

$$y(t) \approx T[0.5x(0) + x(1) + \cdots + x(n-1) + 0.5x(n)]$$

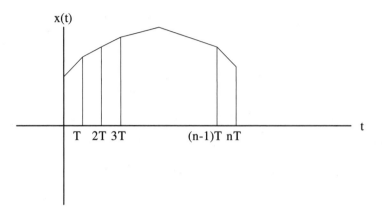

Figure 13.1. Illustration of the trapezoidal integration rule

The last equation is the trapezoidal rule. From the derivation, it can be seen that the rule can be rewritten as shown in Equation 13.3 to give the **trapezoidal digital integration** method.

$$y(t) \approx T[\frac{x(0) + x(1)}{2} + \frac{x(1) + x(2)}{2} + \cdots + \frac{x(n-2) + x(n-1)}{2}$$

$$+ T\frac{x(n-1) + x(n)}{2}$$

$$y(nT) \approx y[(n-1)T] + T\frac{x(n-1) + x(n)}{2} \quad \text{(Equation 13.3)}$$

From this last equation, it can be seen that digital integration can be done continuously on a signal as samples of the signal are received from an ADC by just adding half of the sum of the two most current samples multiplied by the sampling period.

Equation 13.3 is a difference equation, which has a z-transform, as shown in the following equation.

$$Y(z) = z^{-1}Y(z) + \frac{T}{2}[X(z) + z^{-1}X(z)]$$

The z-transfer function can be easily obtained from the z-transformed equation to give the following equation.

$$T(z) = \frac{Y(z)}{X(z)} = \frac{T}{2}\frac{z+1}{z-1}$$

Digital Integration

Thus, using the trapezoidal rule for integration to find an approximation for the total area under a signal given its samples, we have derived the difference equation and next the transfer function used to perform the DSP process of digital integration. The same steps could have been used on Simpson's rule of integration (but only computing the integral every other sample) in order to obtain another DSP process to approximate integration, but the results would have been more complex. Example 13.2 will use the preceding results to perform digital integration on an exponentially decaying signal and compare the result with the integration of the actual signal at the sample times.

Example 13.2. Using the trapezoidal rule to perform digital integration

Problem: Let the signal be given by $5e^{-10t}$ for $t > 0$, with the sampling period $T = 0.02$.

Solution: Using the equation for the trapezoidal rule for digital integration gives the following equations.

$$y(n) = y(n-1) + \frac{0.02}{2}[x(n) + x(n-1)]$$

$$y(n) = y(n-1) + 0.01[5e^{-0.2n} + 5e^{-0.2(n-1)}]$$

$$y(n) = y(n-1) + 0.01(5)[e^{-0.2n} + e^{0.2}e^{-0.2n}]$$

$$y(n) = y(n-1) + 0.05(1 + e^{0.2})e^{-0.2n}$$

$$y(n) = y(n-1) + 0.11107e^{-0.2n} \quad \text{(for } n > 0\text{)}$$

Coding and running this equation in a loop should give a good approximation to the actual integration of the analog signal. Note that for computational convenience the last equation was obtained; but for doing actual DSP integration, the difference equation that was started with would be used, since the actual signal would not be known.

The following table compares the results of using the trapezoidal rule in DSP form versus the actual integral of the analog signal.

TIME	DSP INTEGRATION	ANALOG INTEGRATION
0.00	0.00	0.00
0.02	0.09094	0.09063
0.04	0.16539	0.16484
0.06	0.22635	0.22559
0.08	0.27625	0.27534
0.10	0.31711	0.31606

By comparing the values at the sample times in the last two columns, it can be seen that the DSP version of the trapezoidal rule does a fairly accurate job of finding the area under a decaying exponential signal at every sample time and over one time constant at least.

Now let's look at an even simpler digital integration method called **rectangular** or **Euler integration**. The graphical representation of the rectangular integration method is shown in Figure 13.2. In Example 13.3, we will show that, in general, it is not as good as trapezoidal integration, but it is one of the most widely used digital integration methods used in real-time control systems and digital simulation. The reason will be shown in Section 13.3. Rectangular integration is discussed in Chapter 1, where its unmodified difference equation was given as the following equation.

$$y = y_{-1} + Tx_{-1}$$

Modifying this difference equation into the more standard notation, as was shown in Chapter 5, gives the difference equation as Equation 13.4.

$$y(n) = y(n-1) + Tx(n-1) \qquad \text{(Equation 13.4)}$$

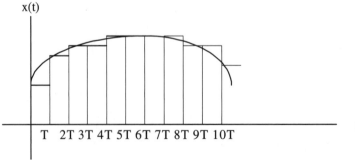

Figure 13.2. Illustration of the rectangular integration method (with quantization error)

Digital Integration

This difference equation is the standard equation for rectangular integration. Instead of averaging the two most current input signal samples, it uses just the previous sample of the input signal. In other words, it assumes the input signal is constant between samples, whereas the trapezoidal rule assumes the signal is a straight line with constant slope between samples, as seen in Figure 13.1. The rectangular method should not be as good as the trapezoidal rule for integration of known signals, unless of course the signal really is a constant value between samples. In Example 13.3, we will use rectangular integration to integrate the same signal as that for trapezoidal integration in Example 13.2, so that the methods can be compared.

Example 13.3. Using rectangular integration to perform digital integration

Problem: Integrate the signal $5e^{-10t}$, $t > 0$ with the sampling period $T = 0.02$ using rectangular integration.

Solution: Using Equation 13.4 for rectangular integration gives the following equations.

$$y(n) = y(n-1) + Tx(n-1)$$

$$y(n) = y(n-1) + T(5e^{-10(n-1)T})$$

$$y(n) = y(n-1) + 0.02[5e^{-0.2(n-1)}]$$

$$y(n) = y(n-1) + 0.12214e^{-0.2n} \quad (\text{for } n > 0)$$

The results of using the last equation are given in the RECT. column in the following table.

TIME	RECT.
0.00	0.0000
0.02	0.100
0.04	0.1819
0.06	0.2489
0.08	0.3038
0.10	0.3487

Comparing the results of Example 13.3 with the results using trapezoidal integration in Example 13.2 shows that rectangular integration gives

Digital Signal Processing

less-accurate results. The only way to improve the accuracy would be to use a shorter sampling period T. Thus, from these last two examples, it would appear that trapezoidal integration is more efficient than rectangular, which is generally the case. Usually several times the sampling frequency and computation rate is required to be used for rectangular integration as opposed to trapezoidal integration. However, we will see in Section 13.3 that rectangular integration is preferred in some cases when digital integration is used to solve differential equations.

13.3 Digital Integration for Differential Equation Solution

In the last section it seems like it was fairly easy to convert the standard rule for trapezoidal integration into a difference equation (recursive) that could perform digital integration on a sampled continuous time signal to give the approximate integral at every sample time. This is very useful when an incoming signal needs to be integrated, such as the output of an inertial accelerometer in order to get the velocity. However, in engineering, one of the most powerful uses of digital integration is in solving differential equations.

Using digital integration to solve even first-order differential equations can lead to strange requirements that can't be met. This is shown again here by starting with a first-order differential equation and then integrating both sides.

$$\frac{dy(t)}{dt} + by(t) = ax(t)$$

$$\int_0^t \frac{dy(t)}{dt} dt + \int_0^t by(t) dt = \int_0^t ax(t) dt$$

Cancelling the dt in the numerator and denominator of the leftmost integral and using the definition of integration as anti-differentiation give the following equations.

$$y(t) - y(0) + b\int_0^t y(t) dt = a\int_0^t x(t) dt$$

$$y(t) = y(0) - b\int_0^t y(t) dt + a\int_0^t x(t) dt$$

As can be seen from the last equation, there is a problem here that would not exist if analog integrators were being used (these are just op amps with capacitors in the negative feedback path). The last equation requires the answer to be used to get the answer. In a computer or DSP chip it will take at least one more computer cycle to use the output in the first integral equation on the right.

The easiest and most obvious solution is to use the previously computed output, but even this brings up another problem, which will be illustrated by the following equations using the trapezoidal digital integration method for both integrals on the right of Equation 13.5.

$$y(t) = y(0) - b\int_0^t y(t)\,dt + a\int_0^t x(t)\,dt$$

$$\int_0^t y(t)\,dt \approx w(n) = w(n-1) + \frac{T}{2}[y(n-1) + y(n-2)]$$

$$\int_0^t x(t)\,dt \approx v(n) = v(n-1) + \frac{T}{2}[x(n) - x(n+1)]$$

$$y(n) = y(0) - bw(n) + av(n) \qquad \text{(Equation 13.5)}$$

In the last equation, w is just the discrete integral of the variable y, and v is just the discrete integral of the variable x. The important item to notice is that using the trapezoidal rule for digital integration has resulted in using older information in the solution for $y(n)$ in the last equation. It was necessary to use the previous value of $y(n)$ for $y(n)$ since $y(n)$ does not exist yet, but we have also used the value of $y(n)$ before that, that is, $y(n-2)$. Using older and older data in recursive equations, which by definition the last equation is, can be very dangerous. This is illustrated in Example 13.4, where trapezoidal integration is used to solve for the output of a first-order differential equation, and then one of the integrals is computed using rectangular integration. Since the output of a first-order linear time-invariant ordinary differential equation is computed, its exact solution is also derived, and the results compared to it.

Example 13.4. Solving a first-order differential equation using digital integration

Problem: Let it be required to compute the output $y(t)$ of the following first-order differential equation using digital integration with the sampling period $T = 0.02$ for a unit step input and no initial conditions.

$$\frac{dy(t)}{dt} + 5y(t) = x(t)$$

Solution: Since this is a simple first-order differential equation, let's first find its solution using Laplace transforms from the student's analog signal processing course. The following equations show the steps and final algebraic solution, where $Y(s)$ and $X(s)$ are the Laplace transforms for $y(t)$ and $x(t)$.

$$sY(s) + 5Y(s) = X(s)$$

$$Y(s)[s+5] = \frac{1}{s}$$

$$Y(s) = \frac{1}{s(s+5)}$$

$$Y(s) = \frac{0.2}{s} + \frac{-0.2}{s+5}$$

$$y(t) = 0.2[1 - e^{-5t}], \; t \geq 0$$

The results of solving this algebraic equation at the sample times is given in the TRUE column in Table 13.1. Now let's use digital integration to compute $y(t)$, using trapezoidal integration as shown in Equation 13.3 and the two equations above it (assuming no initial conditions). These equations then become the following equations.

$$w(n) = w(n-1) + 0.01[y(n-1) + y(n-2)]$$

$$v(n) = v(n-1) + 0.01[x(n) + x(n-1)]$$

$$y(n) = -5w(n) + v(n)$$

Computing the preceding equation recursively, where $w(n)$ and $v(n)$ are just the digital integral values of $y(n)$ and $x(n)$ respectively, gives the values for $y(t)$ at the sample times in the TRAP column. Remember that $x(n)$ is zero for $n < 0$ and 1 otherwise.

Digital Integration

Table 13.1
Comparison of digital integration methods for a first-order equation

TIME	TRUE	TRAP	RECT/TRAP
0.00	0.000	0.000	0.000
0.02	0.019	0.020	0.020
0.04	0.036	0.039	0.038
0.06	0.052	0.056	0.054
0.08	0.066	0.071	0.069
0.10	0.079	0.085	0.082
0.12	0.090	0.097	0.094
0.14	0.101	0.108	0.104
0.16	0.110	0.118	0.114
0.18	0.119	0.126	0.123
0.20	0.126	0.134	0.130

Next, let's replace the trapezoidal integration of $w(n)$ on the right side of Equation 13.5 with the simplest type of digital integration, rectangular integration. This discrete integration is performed by the second equation above Equation 13.5. Now the recursive equations used to compute $y(t)$ at the sample times is given by the following equations for zero initial conditions with a unit step input.

$$w(n) = w(n-1) + 0.02y(n-1)$$

$$v(n) = v(n-1) + 0.01[x(n) + x(n-1)]$$

$$y(n) = -5w(n) + v(n)$$

Computing $y(n)$ at the sample times using the preceding equation gives the values in the RECT/TRAP column of Table 13.1.

From Table 13.1, we see that using the simplest digital integration method, the rectangular, produces better results than a more sophisticated method, the trapezoidal. This is because when using digital integration to solve differential equations, you end up requiring the result to compute the result. For this case, less error usually results by using a single previous value instead of the two previous values required by trapezoidal integration.

Example 13.4 may have given the impression that digital integration is a very hard and laborious way to compute the solution of a differential equation, but it is not. In Example 13.4 a simple first-order linear equation as well as a unit step input were used so that it would be easy to determine the true values analytically. But consider if the differential equation were a higher-order one with a nonconstant input. Then the analytical solution is very complex, but using digital integration takes just a few more steps. If the differential equation is still first order but nonlinear, like the equation examined in Example 13.5, there is no way to use Laplace transforms to find a solution. But the recursive equations using digital integration would only require a small modification to continue to compute the solution to the equation. This will be shown in Example 13.5. With this power, digital integration of differential equations becomes digital simulation of systems described by differential equations!

Example 13.5. Solving a first-order nonlinear differential equation using digital integration

Problem: Determine difference equations to solve the following nonlinear first-order differential equation, using the Euler integration method. Let the input $x(t)$ be a unit step, and let all the initial conditions be zero and the sampling period be 0.02 s.

$$\frac{dy(t)}{dt} + 5\cos(y) = x(t)$$

Solution: Equation 13.5 and the two equations above it are now modified, using Euler integration, to solve the preceding differential equation.

$$w(n) = w(n-1) + 0.02\cos(y(n-1))$$

$$v(n) = v(n-1) + 0.02u(n-1)$$

$$y(n) = y(0) - 5w(n) + v(n)$$

Summary

In Chapter 13 we have looked at digital integration, another use for DSP. Again we saw that difference equations resulted. Only the coefficients

Digital Integration

determine if the difference equation represents digital filtering or digital integration, as was stated in Chapter 1. We also looked at the digital integration of a signal after sampling in Section 13.2. The resulting trapezoidal method was derived from the standard mathematical texts equation for the trapezoidal rule, which was not applicable for giving the result after every sample. Also, a simpler digital integration method called rectangular integration was developed, and the results were compared for integration of a sampled signal. The comparison showed that the trapezoidal method gave more-accurate results.

In Section 13.3 we looked at the problem of solving differential equations by using digital integration and saw there was a basic problem. This problem is that the calculated result is required to get the result, and one solution is to use the previous result to compute the result. Example 13.4 showed that when this is done, rectangular integration gives better results than the usually more accurate trapezoidal method. This illustrates why rectangular integration is used in many cases in place of trapezoidal or much more sophisticated methods.

Since any physical system can be described by differential equations, which may be nonlinear or time varying, the methods of digital integration are used to compute the results at the sample times, which are not now the sample times of an ADC, but are the simulation output times. The simple examples and digital integration methods given here are used for illustrative purposes only, in order to show another aspect of DSP and some of its basic characteristics.

Self-Test

1. Given the following sequence of samples of a signal, use rectangular integration to determine the integral $y(n)$ of the signal at each sample for $T = 0.05$ s. All other samples of the signal are 0.0.

   $x(0) = 0.0, x(1) = 0.3, x(2) = 0.6, x(3) = 0.7, x(4) = 0.8, x(5) = 0.8$

2. Given the sequence of samples in Problem 1, use trapezoidal integration to determine the integral of the signal at each sample up to $y(4)$ for $T = 0.01$ s. All other samples of the signal are 0.0.

3. Given the following differential equation, modify Equation 13.5

to solve for $y(n)$ for $x(t) = u(t)$ with the simulation sampling period $T = 0.1$ s.

$$\frac{dy(t)}{dt} + 2y(t) = x(t)$$

4. In order to compute the equation for $y(n)$ in Problem 3, two auxiliary equations are required. Give the auxiliary equations for trapezoidal integration.

5. Use the results of Problem 4 to write the complete set of difference equations necessary to solve the equations in Problem 3 in their proper order.

6. Modify the answer for Problem 4 to use rectangular integration for both auxiliary equations.

7. Use Mathcad to determine the integral $y(t)$ of the following signal $x(t)$ for every 0.5 s from 0 to 4 s using rectangular integration.

$$x(t) = 0.25t$$

8. Use Mathcad to determine the integral $y(t)$ of the signal in Problem 7 for every 0.5 s from 0 to 4 s using trapezoidal integration.

9. For the following differential equation, use Mathcad to solve for $y(1)$, using trapezoidal integration for $x(t) = u(t)$ for a sampling period $T = 0.1$ s.

$$\frac{dy(t)}{dt} + 3y(t) = 2x(t)$$

Problems

1. Given the following sequence of samples of a signal, use rectangular integration to determine the integral $y(n)$ of the signal at each sample for $T = 0.1$ s. All other samples of the signal are 0.0.

$$x(0) = -0.1, \; x(1) = -0.2, \; x(2) = 0.3, \; x(3) = 0.1, \; x(4) = -0.3$$

2. Given the sequence of samples in Problem 1, use trapezoidal integra-

tion to determine the integral of the signal at each sample up to $y(4)$ for $T = 0.2$ s. All other samples of the signal are 0.0.

$$x(0) = 1.2, \ x(1) = 1.5, \ x(2) = 1.8, \ x(3) = 1.7, \ x(4) = 0.5$$

3. Given the following differential equation, modify Equation 13.5 to solve for $y(n)$ for $x(t) = tu(t)$ with the simulation sampling period $T = 0.02$ s.

$$\frac{dy(t)}{dt} + y(t) = 2x(t)$$

4. In order to compute the equation for $y(n)$ in Problem 3, two auxiliary equations are required. Give the auxiliary equations for rectangular integration.

5. Use the results of Problem 4 to write the complete set of difference equations necessary to solve the equations in Problem 3 in their proper order.

6. Modify the answer for Problem 4 to use trapezoidal integration for both auxiliary equations.

Answers to Self-Test

1. $y(0) = 0.0$, $y(1) = 0.0$, $y(2) = 0.015$, $y(3) = 0.045$, $y(4) = 0.08$, $y(5) = 0.12$, $y(6) = 0.16 = y(7)$ and above.

2. $y(0) = 0.0$, $y(1) = 0.0015$, $y(2) = 0.006$, $y(3) = 0.0125$, $y(4) = 0.02$

3. $y(n) = y(0) - 2w(n) + v(n)$

4. $w(n) = w(n-1) + 0.05y(n-1) + 0.05y(n-2)$
   $v(n) = v(n-1) + 0.05u(n) + 0.05u(n-1)$

5. $w(n) = w(n-1) + 0.05y(n-1) + 0.05y(n-2)$
   $v(n) = v(n-1) + 0.05u(n) + 0.05u(n-1)$
   $y(n) = y(0) - 2w(n) + v(n)$

6. $w(n) = w(n-1) + 0.1y(n-1)$

$$v(n) = v(n-1) + 0.1u(n-1)$$

7. $y(0) = 0.0$, $y(0.5) = 0.0$, $y(2) = 1.75$

8. $y(0) = 0.0$, $y(0.5) = 0.03125$, $y(2) = 2.00$

9. 0.661

appendix A

Laplace Transform Tables

The one-sided Laplace transform $F(s)$ of an analog signal $f(t)$ is defined in the following equation.

$$L[f(t)] = F(s) = \int_0^\infty f(t)e^{-st}\,dt$$

Rather than employ the preceding equation to obtain the Laplace transforms of various signals and properties, tables like Tables A.1 and A.2 are used. Table A.1 gives all the basic signals and their corresponding Laplace transforms, using the preceding equation. Table A.2 gives important properties of all Laplace transforms; in fact, the last property is the reason Laplace transforms are used. This last property will reduce any constant-coefficient time-invariant differential equation into an algebraic equation in the variable s. Using Table A.1 or Partial Fraction Expansion of the resulting expression in s, the resulting time signal for the output can be obtained.

Table A.1 uses the time signal $u(t)$, which is just the continuous time or analog version of the sampled unit step $u(n)$. The analog unit step signal $u(t)$ has a value of 0 for negative time and a value of 1 for nonnegative time values. Any of the time signals in the lefthand column of Table A.1 could be multiplied by the unit step $u(t)$ to give the same Laplace transform in the corresponding righthand column, since the integration over time starts at time 0 and increases.

Example A.1 shows how the equation for the Laplace transform is used to determine the Laplace transform for the time signal that is a constant

Table A.1
Some useful Laplace transform pairs

$f(t)$	$F(s)$
$Au(t)$	$\dfrac{A}{s}$
At	$\dfrac{A}{s^2}$
Ae^{-at}	$\dfrac{A}{s+a}$
Ate^{-at}	$\dfrac{A}{(s+a)^2}$
$A\sin(wt)$	$\dfrac{Aw}{s^2+w^2}$
$A\cos(wt)$	$\dfrac{As}{s^2+w^2}$
$Ae^{-at}\cos(wt)$	$\dfrac{A(s+a)}{(s+a)^2+w^2}$
$Ae^{-at}\sin(wt)$	$\dfrac{Aw}{(s+a)^2+w^2}$

value of A (or $Au[t]$). Example A.2 uses the results of Example A.1 to determine the Laplace transform of Ae^{-at} (or $Ae^{-at}u[t]$).

Example A.1. Using the equation to determine the Laplace transform of a step

Problem: Given the step signal $Au(t)$, use the Laplace transform equation to determine its Laplace transform shown in Table A.1.

Solution: The following mathematical steps show how the Laplace transform equation is used to determine the corresponding Laplace transform of the step of $Au(t)$.

$$L[f(t)] = \int_0^\infty f(t)e^{-st}dt$$

$$L[Au(t)] = \int_0^\infty Au(t)e^{-st}dt$$

Laplace Transform Tables

$$L[A u(t)] = A \int_0^\infty e^{-st} dt$$

$$L[A u(t)] = A \left[\frac{e^{-st}}{-s} \right]_0^\infty$$

$$L[A u(t)] = A \left[\left(\frac{e^{-\infty s}}{-s}\right) - \left(\frac{e^{-0s}}{-s}\right) \right]$$

$$L[A u(t)] = -\left(\frac{A}{-s}\right) + \left(\frac{0}{s}\right) = \frac{A}{s}$$

Example A.2. Using the Laplace equation to determine the Laplace transform of a decaying exponential signal

Problem: The mathematical expression for a decaying exponential signal with a value of A at $t = 0$ and a time constant of $1/a$ s, is given by the following equation. This expression will be used in the Laplace equation to determine the corresponding Laplace transform for the following signal.

$$f(t) = A e^{-at}$$

Solution: The following mathematical steps show how the Laplace equation is used to obtain the Laplace transform of the given decaying exponential signal.

$$L[f(t)] = \int_0^\infty f(t) e^{-st} dt$$

$$L[A e^{-at}] = \int_0^\infty A e^{-at} e^{-st} dt$$

$$L[A e^{-at}] = A \int_0^\infty e^{-(s+a)t} dt$$

Note: The last integral is the same as that in Example A.1, except s is replaced by $s + a$. Thus the result is the same as in Example A.1, except that s is replaced by $s + a$, as shown in the following equation.

Digital Signal Processing

$$L[Ae^{-at}] = \frac{A}{s+a}$$

The last property in Table A.2 includes the initial condition of the signal $f(t)$. Usually in analog or digital filtering the initial condition is ignored, so the term $f(0)$ is usually dropped. In Example A.3, the shifting property is derived using the Laplace equation. This property is used in determining some of the characteristics of z-transforms in the text.

Example A.3. Determining the effect of a time shift of a signal on its Laplace transform

Problem: The original time signal is $f(t)$; if it is delayed by T s, then it is written as $f(t - T)$. We will use the Laplace equation to determine the Laplace transform of the shifted time signal $f(t - T)$ given that the Laplace transform of the unshifted signal is $F(s)$.

Solution: The following mathematical equations show the steps used to derive the Laplace transform of $f(t - T)$, where $m = t - T$ is used and $f(t)$ or $f(m)$ is assumed zero for negative arguments.

$$L[f(t - T)] = \int_0^\infty f(t - T)e^{-st}dt$$

$$L[f(t - T)] = \int_{-T}^\infty f(m)e^{-s(m+T)}dm$$

$$L[f(t - T)] = e^{-sT}\int_{-T}^\infty f(m)e^{-sm}dm$$

$$L[f(t - T)] = e^{-sT}F(s)$$

Table A.2
Some useful Laplace transform properties

$f(t) + g(t)$	$F(s) + G(s)$
$Af(t)$	$AF(s)$
$f(t)*g(t)$	$\neq F(s) * G(s)$
$f(t - T)$	$e^{-sT}F(s)$
$\dfrac{df(t)}{dt}$	$sF(s) - f(0)$

appendix B

Entering a Mathcad Program

Appendix B shows the procedure for writing a Mathcad program. By following the comments on the right, the equations on the left can be obtained. The result is the program and resulting plot given in Example 6.4. The comments also briefly give the significance of each equation.

$n := 1 .. 50$ Type: n:1;50   to compute 50 values to plot.

$w_n := 10 \cdot n$ Type: w[n:10*n   to compute the frequency at each value from 10 to 500 rad/s.

$T := 0.01$ Type: T:0.01   to specify the sampling period as 0.01 seconds.

$z_n := \cos(w_n \cdot T) + \sin(w_n \cdot T) \cdot j$ Type: z[n:cos(w[n *T)+sin(w[n T) *1j   to convert z to jw for each value of z used. The spaces must also be typed.

$gain_n := \left| 0.5 + 0.5 \cdot (z_n)^{-1} \right|$ Type: gain[n:|0.5+0.5z[n^-1   to compute the gain for each value of n for the given T(z) inside the absolute value signs.

$gaindB_n := 20 \cdot \log(gain_n)$ Type: gaindB[n:20*log(gain[n)   to convert gain to gain in decibels.

Digital Signal Processing

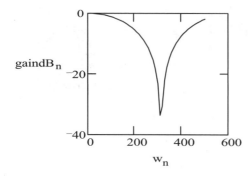

Type: @ to display graph template.

Type: gaindB[n and w[n on top of graph template place holders (small square boxes) at places they appear to left.

Double click inside of graph to display options to change like color, axes scaling, dots or lines, then click outside of graph area after okay.

$w_{20} = 200$

Type w[20= to see the value for the frequency at n = 20.

$gaindB_{20} = -5.347$

Type gaindB[20= to see the gain in dB at n = 20

Index

A

a coefficient, determining change in zero corner frequency, 138–139
ADC (Analog to digital converter)
    computing for 8 bit, 194
    effects of number of bits, 191–194
    signal sampling/conversion and, 3
Aliasing, 15
Analog filter
    approximations, 82–84
    corner, predicting change in, 135
Analog system, unstable, 112–113
Analog to digital converter (ADC)
    computing for 8 bit, 194
    number of bits, effects of, 191–194
    signal sampling/conversion and, 3
Anti-aliasing filters, 16–17
    determining requirements for, **17**
    lowpass digital filter and, **17**
Approximations, analog filter, 82–84

B

b coefficient
    finding, for specified pole corner frequency accuracy, 140–141
    precision of, two-pole IIR filter, 136–137
Bandpass filter
    defined, 31
    finding first two coefficients with non-negative subscripts for, 153–154
    specification, 31–33
Bandstop filter, specification, 33–35
Bartlett window
    computing/using, on lowpass FIR filter, 168–169
    FIR filter(s) and, 167–169
Bilinear Transform (BLT) filter, 94–102
    find the, IIR filter for first-order analog highpass filter, 95–96
BLT (Bilinear transform filter), 94–102
    finding the, IIR filter for first-order analog highpass filter, 95–96
    IIR filters, computing for first-order lowpass analog filters, 91–93
Butterworth approximation, 82

C

Canonical form of IIR filter, 197–199
Cascade form of IIR filter, 199–200
Casual coefficients, FIR filters and, 156–160
Cauer approximation, 84
Chebyshev approximation, 82
Coefficient word length effects, DSP, 139–141

Coefficients, arbitrary, DSP equation and, 57
Cosine signal
 frequency of, increasing, 12, 13
 periodical sampling of, 9–11
Cosine wave, signal for, 10

D

DAC (Digital to analog converter), signal sampling/conversion and, 3
Data
 smoothing, 158–159
 using noncasual filter on, 157–158
Design considerations
 effects of, analog to digital converter number of bits, 191–194
 filters, realization forms for, 196–201
 FIR versus IIR filters, 188–191
 fixed point math versus floating point, for DSP chip, 194–195
Difference equation, 58
 finding for T(z), 61–62
 z-transform of, 59
Differential equations
 digital integration and, 214–218
  solving first-order, 215–217
  solving first-order nonlinear, 218–219
Digital amplifier, equation for, 3
Digital filter
 determining characteristics of, using unit circle, 122–123
 DSP stability and, 111–127
  z-plane unit circle and, 113–120
 finding transfer function for, 60–61
 specifications, 25–37
  alternate graphical, 35–36
  bandpass digital filter, 31–33
  bandstop digital filter, 33–35
  dB, 25–27
  filter gain, 25–27
  graphical filter, 25–27
  highpass digital filter, 30–31
  loss, 25–27
  lowpass digital filter, 28–29
 transfer function of, 59–62
Digital filtering
 example of, **5**
 reason for, 1
Digital frequency, described, 35

Digital integration
 described, 207–209
 differential equations and, 214–218
  solving first-order, 215–217
  solving first-order nonlinear, 218–219
 example of, **3, 4**
 of known signals, 209–214
 using rectangular integration, 213
 using trapezoidal rule, 211–212
Digital lowpass, filtering, 4
Digital representation, precision of, 130
Digital signal processing (DSP)
 coefficient word length effects, 139–141
 defined, 1, 2
 equation for, 5
  with arbitrary coefficients, 58
  modified, 55–58
  questions raised by, 6–7
 equivalence of output for sinusoids, 14
 examples of, 2–5
 flowchart of, **6**
 frequency response, 70–74
 historical perspective of, 1–2
 output periodicity by mathematical means, 18–20
 stability, 111–113
 system frequency response, periodicity of, 11–14
Digital to analog converter (DAC), signal sampling/conversion and, 3
Direct form of IIR filter, 196
DSP. *See* Digital signal processing

E

Euler equation, 67–69
 using, 69
Euler integration, 212–214

F

Filter approximations, analog, 82–84
Filter coefficient precision
 computing coefficient word length effects, 139–141
 development of equations, for precision effects, 131–139
 numeric precision and, 129–131
Filter corner frequency, stability of, determining using relation-

Index

ship of z plane to the s plane, 123–124
Filter gain, defined, 26
Filter specification, jump elimination in, 176–178
Filters, realization forms for, 196–201
Finite Impulse Response Filter. *See* FIR filter(s)
FIR filter(s)
   basic solutions, use of, 154–156
   coefficient equation, 147–150
      basic solutions of them, 150–154
   coefficients,
      casual (real time)/noncasual, 156–160
      for a LPF, 149–150
      with nonnegative subscripts, 153–154
   described, 146–147
   design/check with Mathcad, 155–156
   Gibbs effect and, 165–167
   non-windowing approaches to, 175–183
   transversal form of, 200–201
   versus IIR filters, 188–191
   windows for, 165–186
      Bartlett, 167–169
      Hamming, 172–175
      von Hann, 170–172
First-order highpass filter, finding with BLT method, 95–96
Fixed point math versus floating point, for DSP chip, 194–195
Fixed point number
   computing range of values by, 195
   representations defined, 194
Floating point math versus fixed point, for DSP chip, 194–195
Floating point number
   computing range of values by, 195
   representations, 194
"For" loop, 4
Frequency
   filter corner, determine stability of using relationship of z plane to the s plane, 123–124
   plotting, **27**
Frequency response, 67–80
   computing, 70–74
   digital signal processing, computing, 70–74
      periodicity of, 11–14
   Euler equation and, 67–69
   frequency scaling and, 699–670

Frequency scaling, 69–70

G

Gain dB
   calculating/plotting using Mathcad, 73
   computing from gain, **27**
   defined, 26
   finding one value of input frequency, 72–73
   one frequency calculation, filter with a and b coefficients, 74
Gibbs effect, 165–167
   non-windowing approaches, 175–183
   windowing approaches to, 167–175
      Bartlett window, 167–170
      Hamming window, 172–175
      Hann window, 170–172
$g_{p\text{max}}$, defined, 28
$g_{p\text{min}}$, defined, 28
$g_{s\text{max}}$, defined, 28

H

Hamming window, FIR filters and, 172–175
Hann window, FIR filters and, 170–172
Highpass filter
   defined, 30
   first-order, finding with BLT method, 95–96
   specification, 30–31

I

IIR filter(s)
   analog filter approximations and, 82–84
   bilinear transform (BLT) method, 94–102
      for first-order analog highpass filter, 95–96
   canonical form of, 197–199
   cascade form of, 199–200
   computing the BLT, for first-order lowpass analog filters, 131–133
   described, 81–82
   direct form of, 196
   impulse invariant, 84–91
      finding for a more complex analog filter, 89–91

finding to approximate a first-order analog filter, 87–89
precision of b coefficient of, 136–137
stability of, using z-plane to determine, 117–118
step invariant,
  to approximate a first-order lowpass analog filter, 93
  method, 91–93
versus FIR filters, 188–191
Impulse function, z-transform, for short sampled signal description, 42
Impulse invariant IIR filter, 84–91
  analog filter approximations and, 82–84
  bilinear transform (BLT) method, 94–102
  canonical form of, 197–199
  cascade form of, 199–200
  computing the BLT, for first-order lowpass analog filters, 131–133
  described, 81–82
  direct form of, 196
  finding,
    to approximate a first-order analog filter, 87–89
    for a more complex analog filter, 89–91
  precision of b coefficient of, 136–137
  stability of, using z-plane to determine, 117–118
  step invariant method, 91–93
Infinite Impulse Response (IIR) filter. *See* IIR filter
Input frequency, finding gain for one value of, 72–73
Input signal, lowpass filtering of, 4
Integral, of signal/function, 4
Inverse Chebyshev approximation, 83–84

J

Jump elimination, in filter specification, 176–178

L

Laplace transform tables, 223–226
Loss, defined, 27
Lowpass filter
  anti-aliasing filter and, **17**
  calculating and using,
    Bartlett window and, 168–169
    Hamming window coefficients on, 174–175
    Hann window coefficients on, 170–172
    with a transition region specified, 177–178
  casual, smoothing data with, 158–159
  finding,
    step invariant IIR filter to approximate first-order, 93
    transfer function of, 59–61
  FIR coefficients with nonnegative subscripts for, 150
  magnitude spectrum, **15**
  purpose of, 12
  specification, 28–29
Lowpass filtering, of input signal, 4
Lowpass filters, computing first order, 91–93

M

Magnitude transfer function, defined, 26
Mathcad program
  calculate/plot gain with, 73
  design/check FIR LPF design with, 155–156
  entering, 227–228

N

Noncasual coefficients, FIR filters and, 156–160
Numeric precision, 129–131
  of digital representation, 130
Nyquist criterion, defined, 15
Nyquist limit
  defined, 15
  output periodicity by mathematical means, 18–20

P

Parks-McClellan method, 178–180
Partial Fraction Expansion (PFE), 200
Passband, defined, 28
PFE (Partial Fraction Expansion), 200
Pole corner frequency accuracy, finding for b coefficient, 140–141

Index

Precision
  coefficient, illustrating effect of sampling period on required, 136
  computer numeric, 129–131
  determining number of bits to ensure, 131
  equation development, 131–139

Q

Quantization noise, described, 191

R

Real time coefficients, FIR filters and, 156–160
Realization forms, for filters, 196–201
Rectangular integration, 212–214
  digital integration with, 213
Representations
  fixed point, 194
  floating point number, 194

S

Sampled signal, z-transform of, 40–41
Sampling, cosine signal, 9–11
Scaled frequency, 69–70
  described, 35
  stopband filter specification using, 35–36
Shifting property, defined, 46
Signal sampling
  aliasing, 15–16
  anti-aliasing filters, 16–17
  Nyquist limit, 15–16, 18–20
  output periodicity by mathematical means, 18–20
  periodic, of cosine signal, 9–11
  system frequency response, 11–14
Signal to noise ratio (SNR)
  computing for 8 bit ADC, 194
  described, 191
Signals, known, digital integration of, 209–214
Simpson's Rule, 207–208
  finding integral of sample data with, 208
Sinusoids, DSP equivalence of output for, 14
SNR (Signal to noise ratio)
  computing for 8 bit ADC, 194
  described, 191
Stability
  DSP, 111–113
  z-plane unit circle and, 113–120
Step invariant IIR filter, 91–93
  to approximate a first-order lowpass analog filter, 93
Stopband, defined, 28
Stopband filter, specification for, 35–36
Stored data, using noncasual filter on, 157–158
System frequency response, digital signal processing (DSP), periodicity of, 11–14

T

Transfer function
  of digital filter, 59–62
  finding for,
    complex digital filter, 60–61
    lowpass filter, 59–61
Transition band, defined, 28
Transversal for of FIR filters, 200–201
Trapezoidal digital integration method, 210
Trapezoidal rule, 207–208
  digital integration using, 211–212
  finding integral of sample data with, 208
$T(z)$
  finding difference equation for, 61–62
  using pole location to determine stability of, 118–120

U

Uniform distribution, defined, 192
Unit circle, determining digital filter characteristics using, 122–123
Unstable
  analog system, 112–113
  described, 111–113

V

von Hann window, FIR filters and, 170–172

W

W_f, defined, 28
"While" loop, 4
Windowing, defined, 166
W_p, defined, 28
W_s, defined, 28

Z

Zero corner frequency, determining change in, 138–139
Z-plane
   pole locations,
      to determine stability of an IIR filter, 117–118
      to determine stability of an unfactored T(z), 118–120
   properties of, 120–124
   unit circle, stability and, 113–120
z-Transforms
   defined, 40
   derivation of,
      necessary pairs of, 43–46
      using Algebra, 46–49
   of a difference equation, 59
   of DSP equation, 55–66
   with few sample values, 41
   need for, 39–40
   signals of, **44**
   uses of, 40–43
   using impulse function for short sampled signal description, 42
   writing the,
      equation with shifted functions, 47–48
      sampled signal delayed by 3 sample periods, 47